Data-Centric Machine Learning with Python

The ultimate guide to engineering and deploying high-quality models based on good data

Jonas Christensen

Nakul Bajaj

Manmohan Gosada

Data-Centric Machine Learning with Python

Group Product Manager: Niranjan Naikwadi
Publishing Product Manager: Sanjana Gupta
Book Project Managers: Farheen Fathima and Aparna Ravikumar Nair
Content Development Editor: Manikandan Kurup
Technical Editor: Kavyashree K S
Copy Editor: Safis Editing
Proofreader: Safis Editing
Indexer: Hemangini Bari
Production Designer: Ponraj Dhandapani
DevRel Marketing Coordinator: Vinishka Kalra

First published: February 2024

Production reference: 1220224

Published by Packt Publishing Ltd.
Grosvenor House
11 St Paul's Square
Birmingham
B3 1RB, UK.

ISBN 978-1-80461-812-7

www.packtpub.com

Foreword

In the current era of ubiquitous data, **machine learning (ML)**, and **artificial intelligence (AI)**, we find the pace of many things is accelerating – innovation, technologies, new products, business processes, and consumer applications. This is a marvelous time for those who work and play, those who do research and development, and those who dream and build within this sphere. You might say that this is the result of thousands of years of human development. How is that?

The present explosion of global interest and scrutiny, dreams and start-ups, products and services, and tension between risk and reward that are associated with AI is unsurprisingly coincident with the exploitation of large and rich data collections in every facet and dimension of human endeavor. AI is not new, as it has been around for several decades, although it has become most prominently visible to the general public within the last decade.

AI is built on ML (in various forms), which is essentially a branch of mathematical algorithms and methods that discover and learn patterns in data. ML has been a very active research area for decades, rising to prominence among researchers and especially practitioners on a more accelerated curve than AI. Supporting, driving, and inspiring all of this is our human curiosity and our drive to understand, describe, model, predict, and optimize (when possible) what we see and observe in our world (in fact, in the universe).

The human "see and observe" process is fundamentally data collection – the acquisition and curation of collections of "facts" – acquired through sensors, first with our human senses (beginning thousands of years ago), progressing systematically through the ages with increasingly sophisticated collection and recording methods and technologies. The more sophisticated, efficient, and comprehensive those methods and technologies became, the greater our data collections grew. And so, here we are – living in an era of massive datasets that we are driven to explore, innovate with, exploit, create value from, and monetize.

Perhaps it is the "oldness" of data collection, in contrast to the novelty and "rock star" status of ML and AI (plus the actual fun that some of us model-builders and model-users have in building and deploying ML and AI models), that has led to the model-centric focus and strategies in current practice. While that may be a reasonable way of approaching the incredible opportunities that ML and AI systems offer, we must not drift from the foundational principles upon which these systems are established, fueled, and nourished. That foundation is data.

Any methodology that minimizes our attention to data strategy (including its many dimensions, such as data quality, bias, accuracy, validity, freshness, consistency, labeling, completeness, accessibility, usability, relevance, security, privacy, literacy, and fluency) is a multi-step recipe for potential failure or inappropriateness of the ML and AI models upon which the data is built. Neglecting these many dimensions of data strategy is a risky business.

One way to think of "data neglect" is the following. We give a lot more time and attention to the outcomes and goals (destinations) of our model-building ML and AI activities (such as improved accuracy and speed) than to the broad spectrum of data strategy dimensions. Furthermore, especially for model-builders, the journey (building, validating, and improving our ML and AI models) is where we love to spend our time and expertise. That's analogous to saying that the fuel that powers your automobile toward your vacation paradise receives less of your attention than the journey and the destination. Nevertheless, you are not going anywhere without the right fuel. Data is the fuel, and the "rightness" of the data can be measured in many dimensions (as listed previously). Understanding this perspective on data-centricity leads us to the key motivations, fundamental thesis, important messages, and comprehensive selection of materials presented in this essential book.

Data-centricity may seem like a paradox. It's a bit like the chicken and egg paradox (which one came first?). With a data-centric mindset, it appears that you are being cautioned to think a bit less about the ultimate goal of the thing that you are building and a bit more about the materials that go into the building. But shouldn't the opposite be true? Of course, the end goal is always paramount, as we often hear statements such as these: "*Start with the end in mind,*" "*Be mission-led,*" and "*Know and be guided by your north star.*"

Therefore, instead of dwelling on the paradox, a better data-centric mindset is to focus on those who build models from data. Consequently, this book highlights the importance of diverse data teams, collaboration, "many eyes on the data," data democratization, data champions, and data science as a team sport. As Antoine de Saint-Exupéry (a French writer, poet, journalist, and pioneering aviator) once exhorted others: "*If you want to build a ship, don't drum up the [people] to gather wood, divide the work, and give orders. Instead, teach them to yearn for the vast and endless sea.*"

Here is my final exhortation to you who are now reading this book – read onward and dream upward, with the goals of your modeling quest strategically in mind, and with vast seas of quality data confidently in hand.

Kirk D. Borne, Ph.D.

Advisor, Mentor, Trainer, Influencer, Astrophysicist

Founder and Owner, Data Leadership Group LLC

Contributors

About the authors

Jonas Christensen has spent his career leading analytics and data science functions across multiple industries. He is an international keynote speaker, postgraduate educator, and advisor in the fields of data science, analytics leadership, and machine learning. He is also the co-author of *Demystifying AI for the Enterprise* and the host of the *Leaders of Analytics* podcast.

Nakul Bajaj is a data scientist, MLOps engineer, educator, and mentor, helping students and junior engineers navigate their data journey. He has a strong passion for MLOps, with a focus on reducing complexity and delivering value from machine learning use cases in business and healthcare.

Manmohan Gosada is a seasoned professional with a proven track record in the dynamic field of data science. With a comprehensive background spanning various data science functions and industries, Manmohan has emerged as a leader in driving innovation and delivering impactful solutions. He has successfully led large-scale data science projects, leveraging cutting-edge technologies to implement transformative products. With a postgraduate degree, he is not only well-versed in the theoretical foundations of data science but is also passionate about sharing insights and knowledge. A captivating speaker, he engages audiences with a blend of expertise and enthusiasm, demystifying complex concepts in the world of data science.

About the reviewers

Valentine Shkulov is a renowned visiting lecturer at a top tech university, where he seamlessly melds academia with real-world expertise as a distinguished data scientist in Fintech, e-commerce, and e-sport. His ingenuity in crafting ML-driven solutions has transformed businesses, from budding start-ups to tech giants. Valentine excels at introducing AI innovations and refining current systems, ensuring they profoundly influence vital business metrics. His passion for navigating product challenges has established him as a pioneer in leveraging ML to elevate businesses.

I am grateful to my friends for their support and to my colleagues for opportunities, each shaping my journey in invaluable ways. Thank you all.

Indraneel Chakraborty is a senior developer currently working in the biomedical software industry. He excels in building web applications and data solutions on the cloud. Proficient in Python and R, and recently venturing into Golang, he has a diverse skill set. Indraneel's journey spans B2B SaaS start-ups, academic research with publications in respected journals, and active contributions to open source projects. He has explored leveraging LLMs for high-value solutions. He also volunteers as a maintainer for open source coding courses and reviews technical book publications. Passionate about tech and data, Indraneel's enthusiasm lies in exploring new technology stacks, honing his cloud engineering skills, and advancing his expertise in the industry.

Table of Contents

Part 2: The Building Blocks of Data-Centric ML

3

4

Part 3: Technical Approaches to Better Data

5

6

Techniques for Programmatic Labeling in Machine Learning   147

7

Using Synthetic Data in Data-Centric Machine Learning   191

8

Techniques for Identifying and Removing Bias 243

9

Dealing with Edge Cases and Rare Events in Machine Learning 293

Part 4: Getting Started with Data-Centric ML

10

Preface

If you're reading this, you've taken the first steps on a pioneering journey to building and implementing machine learning models that are more robust, accurate, fairer, less biased, and easier to explain.

This is a big claim, we know. We are comfortable making it, however, on the basis of the huge and relatively untapped potential we see in the data-centric approach to machine learning development.

Why do we consider data-centric machine learning pioneering?

It may seem obvious that improving data quality will lead to more predictive models. However, machine learning research to date has mainly focused on evolving the various algorithms and tools to build and tune models.

As a result, we have available at our fingertips a vast array of machine learning algorithms, tools, and techniques that can give us great models at a low cost, given the right quality and volume of input data.

Model architectures are largely a solved problem in most situations. What data scientists, and the organizations they work in, typically lack are best-practice frameworks, tools, and techniques for improving data quality.

Data-centric machine learning builds on the predominant model-centric approach to model development by exploiting the big opportunities that lie in better input data.

Putting a bigger emphasis on data collection and engineering requires us to streamline our processes for collecting quality data and invent new techniques for engineering datasets that provide more signals with much less data.

Many of the techniques and examples you will learn about in this book are based on cutting-edge research and the application of modern practices to collecting, engineering, and synthetically generating great datasets.

Data-centric machine learning also necessitates a much stronger collaboration between data scientists, subject-matter experts, and data labelers. As you will learn throughout this book, data-centricity typically starts with humans collecting and labeling data in a way that serves operational and data science needs.

In many organizations, it is uncommon to collect data for machine learning purposes specifically. A more systematic approach to collecting and labeling data for data science will not only lead to better data but also bring together the thinking and creativity of subject-matter experts and data scientists. This positive feedback loop between different kinds of domain experts creates new opportunities for ideas to flourish far beyond the scope of individual machine learning projects.

Why do we claim that data-centric models will be better than their model-centric counterparts in almost every aspect?

Think of any high-quality consumer product you use regularly. It may be your computer, the car you drive, the chair you sit on, or something else that has required some level of design and engineering.

What makes it high-quality?

Design and functionality have a lot to do with it, but unless the product is made of quality materials, it will not work as intended or it may break altogether. Something is only high-quality if it works as intended, and does so consistently.

The same goes for machine learning models. By systematically improving data quality – our building materials – we are able to build models that are more predictive, robust, and interpretable.

We have written this book to give you, our readers, the most important background knowledge, tools, techniques, and applied examples needed to implement data-centric machine learning and take part in the next phase of the AI revolution.

In the technical chapters of this book, we will show you how to apply the principles of data-centric machine learning to real datasets, using Python. The techniques and applied examples we explore will provide you with a toolbox to systematically and programmatically collect, clean, augment, and label data, as well as to identify and remove unwanted bias.

At the end of this book, you will have a strong appreciation for the building blocks and best-practice approaches of data-centric machine learning.

Don't just take our word for it. Let's explore data-centric machine learning in depth.

Who this book is for

This book is for data science professionals and machine learning enthusiasts wanting to understand what data-centricity is, its benefits over a model-centric approach, and how to apply a best-practice data-centric approach to their work.

This book is also for other data professionals and senior leaders wanting to explore tools and techniques to improve data quality and how to create opportunities for "small data" ML/AI in their organizations.

What this book covers

Chapter 1, Exploring Data-Centric Machine Learning, contains a comprehensive definition of data-centric machine learning and draws contrasts with its counterpart, model-centricity. We use practical examples to compare empirical performance and illustrate key differences between these two methodologies.

Chapter 2, From Model-Centric to Data-Centric – ML's Evolution, takes you on a journey through the evolution of AI and ML toward a model-centric approach, highlighting the untapped potential in improving data quality over model tuning. We also debunk the "big data" myth, showing how shifting to "good data" can democratize ML solutions. Get ready for a fresh perspective on the power of data in ML.

Chapter 3, Principles of Data-Centric ML, sets the stage for your journey into the heart of data-centric ML by outlining the four key principles of data-centric ML. These principles offer crucial context – the *why* – before we delve into the specific methods and approaches linked to each principle – the *what* – in the ensuing chapters.

Chapter 4, Data Labeling Is a Collaborative Process, explores the pivotal role of subject-matter expertise, trained labelers, and clear instructions in ML development. In this chapter, you will learn about the human-centric nature of data labeling and acquire strategies to enhance it to reduce bias, increase consistency, and build richer datasets.

Chapter 5, Techniques for Data Cleaning, explores the six crucial aspects of data quality and showcases various techniques for cleaning data, a vital process for enhancing data quality by rectifying errors. We illustrate why questioning and systematically improving data quality is crucial for reliable machine learning systems, all while teaching you essential data cleaning skills.

Chapter 6, Techniques for Programmatic Labeling in Machine Learning, focuses on programmatic labeling techniques for boosting data quality and signal strength. We go through the pros and cons of programmatic labeling and provide practical examples of how to execute and validate these techniques.

Chapter 7, Using Synthetic Data in Data-Centric Machine Learning, introduces synthetic data as an efficient and cost-effective method for overcoming the limitations of traditional data collection and labeling. In this chapter, you will learn what synthetic data is, how it's used to improve models, the techniques to generate it, and its risks and challenges.

Chapter 8, Techniques for Identifying and Removing Bias, focuses on the problem of bias in the way we collect data, apply data and models to a problem, and the inherent human bias captured in many datasets. We will go through data-centric techniques for identifying and correcting biases in an ethical manner.

Chapter 9, Dealing with Edge Cases and Rare Events in Machine Learning, explains the process of detecting rare events in ML. We explore various methods and techniques, discuss the importance of evaluation metrics, and illustrate the wide-ranging impacts of identifying rare events.

Chapter 10, Kick-Starting Your Journey in Data-Centric Machine Learning, sheds light on the technical and non-technical challenges you might face during model development and deployment. This final chapter shows you how a data-centric approach can help you overcome these challenges, opening up big opportunities for growth and wider use of machine learning in your organization.

To get the most out of this book

To extract the maximum value from this book, prior exposure to machine learning concepts, foundational knowledge of statistical methods, and familiarity with Python programming will be highly beneficial. The book is tailored for those with familiarity with the machine learning process and a desire to delve deeper into the world of data-centric machine learning and artificial intelligence.

Software/hardware covered in the book	Operating system requirements
Python 3	Windows, macOS, or Linux

If you are using the digital version of this book, we advise you to type the code yourself or access the code from the book's GitHub repository (a link is available in the next section). Doing so will help you avoid any potential errors related to the copying and pasting of code.

Download the example code files

You can download the example code files for this book from GitHub at `https://github.com/PacktPublishing/Data-Centric-Machine-Learning-with-Python`. If there's an update to the code, it will be updated in the GitHub repository.

We also have other code bundles from our rich catalog of books and videos available at `https://github.com/PacktPublishing/`. Check them out!

Conventions used

There are a number of text conventions used throughout this book.

`Code in text`: Indicates code words in text, database table names, folder names, filenames, file extensions, pathnames, dummy URLs, user input, and Twitter handles. Here is an example: "We will call the `loan_dataset.csv` file and will save it in the same directory, from where we will run this example."

A block of code is set as follows:

```
import pandas as pd
import os
FILENAME = "./loan_dataset.csv"
DATA_URL = "http://archive.ics.uci.edu/ml/machine-learning-
databases/00350/default%20of%20credit%20card%20clients.xls"
```

Bold: Indicates a new term, an important word, or words that you see onscreen. For instance, words in menus or dialog boxes appear in **bold**. Here is an example: "Biases in machine learning can take many forms, hence we categorized these biases into two main types, **easy to identify** biases and **difficult to identify** biases."

> **Tips or important notes**
> Appear like this.

Get in touch

Feedback from our readers is always welcome.

General feedback: If you have questions about any aspect of this book, email us at customercare@packtpub.com and mention the book title in the subject of your message.

Errata: Although we have taken every care to ensure the accuracy of our content, mistakes do happen. If you have found a mistake in this book, we would be grateful if you would report this to us. Please visit www.packtpub.com/support/errata and fill in the form.

Piracy: If you come across any illegal copies of our works in any form on the internet, we would be grateful if you would provide us with the location address or website name. Please contact us at copyright@packt.com with a link to the material.

If you are interested in becoming an author: If there is a topic that you have expertise in and you are interested in either writing or contributing to a book, please visit authors.packtpub.com.

Share Your Thoughts

Once you've read *Data-Centric Machine Learning with Python,* we'd love to hear your thoughts! Scan the QR code below to go straight to the Amazon review page for this book and share your feedback.

https://packt.link/r/1-804-61812-8

Your review is important to us and the tech community and will help us make sure we're delivering excellent quality content.

Download a free PDF copy of this book

Thanks for purchasing this book!

Do you like to read on the go but are unable to carry your print books everywhere?

Is your eBook purchase not compatible with the device of your choice?

Don't worry, now with every Packt book you get a DRM-free PDF version of that book at no cost.

Read anywhere, any place, on any device. Search, copy, and paste code from your favorite technical books directly into your application.

The perks don't stop there, you can get exclusive access to discounts, newsletters, and great free content in your inbox daily

Follow these simple steps to get the benefits:

1. Scan the QR code or visit the link below

https://packt.link/free-ebook/9781804618127

2. Submit your proof of purchase
3. That's it! We'll send your free PDF and other benefits to your email directly

Part 1: What Data-Centric Machine Learning Is and Why We Need It

In this part, we take a deep dive into data-centric machine learning, contrasting it with model-centric approaches. We use real-life examples to illustrate their differences and explore the evolution of AI and ML toward a data-centric perspective. We also dispel the myth of "big data," highlighting the importance of quality over quantity, and the potential for democratizing ML solutions. Prepare for a fresh perspective on the transformative power of data in ML.

This part has the following chapters:

- *Chapter 1, Exploring Data-Centric Machine Learning*
- *Chapter 2, From Model-Centric to Data-Centric – ML's Evolution*

1

Exploring Data-Centric Machine Learning

This chapter provides a foundational understanding of what data-centric **machine learning** (ML) is. We will also contrast data centricity with model centricity and compare the performance of the two approaches, using practical examples to illustrate key points. Through these practical examples, you will gain a strong appreciation for the potential of data centricity.

In this chapter, we will cover the following main topics:

- Understanding data-centric ML
- Data-centric versus model-centric ML
- The importance of quality data in ML

Understanding data-centric ML

Data-centric ML is the discipline of systematically engineering the data used to build ML and **artificial intelligence** (AI) systems[1].

The data-centric AI and ML movement is grounded in the philosophy that data quality is more important than data volume when it comes to building highly informative models. Put another way, it is possible to achieve more with a small but high-quality dataset than with a large but noisy dataset. For most ML use cases, it is not feasible to build models based on very large datasets, say millions of observations, simply because the volume of data doesn't exist. In other words, the potential use of ML as a tool to solve certain problems is often ignored on the basis that the available dataset is too small.

But what if we can use ML to solve problems based on much smaller datasets, even down to less than 100 observations? This is one challenge the data-centric movement is attempting to solve through systematic data collection and engineering.

For most ML use cases, the algorithm you need already exists. The quality of your input data (x) and your dependent variable labels (y) is what makes the difference. The traditional response to dealing with noise in a dataset is to get as much data as possible to average out anomalies. Data centricity tries to improve the signal in the data such that more data is not needed.

It's important to note that data centricity marks the next frontier for larger data solutions too. No matter how big or small your dataset is, it is the foundational ingredient in your ML solution. Let's take a closer look at the different aspects of data-centric ML.

The origins of data centricity

The push toward a more data-centric approach to ML development has been spearheaded by famous data science pioneer, Dr. Andrew Ng.

Dr. Ng is the co-founder of the massive open online course platform Coursera and an adjunct professor in computer science at Stanford University. He is also the founder and CEO of DeepLearning.AI, an education company, and Landing AI[2], an AI-driven visual inspection platform for manufacturing. He previously worked as chief scientist at Baidu and was the founding lead of the Google Brain team. His Coursera courses on various ML topics have been completed by millions of students worldwide.

Dr. Ng and his team at Landing AI build complex ML solutions, such as computer vision systems used to inspect manufacturing quality. Through this work, they observed that the following characteristics are typical of most ML opportunities[3]:

- The majority of potential ML use cases rely on datasets smaller than 10,000 observations. It is often very difficult or impossible to add more data to reduce the effects of noise, so improving data quality is essential to these use cases.

- Even in very large datasets, subsets of the data will exhibit the behavior of a small dataset. As an example, Google's search engine generates billions of searches every day, but 95% of the searches are based on keyword combinations that occur fewer than 10 times per month (in the US). 15% of daily keyword combinations have never been searched before[4].

- When the dataset is small, it is typically faster and easier to identify and remove noise in the data than it is to collect more data. For example, if a dataset of 500 observations has 10% mislabeled observations, it is usually easier to improve the data quality on this existing data than it is to collect a new set of observations.

- ML solutions are commonly built on pretrained models and packages, with minimal tweaking or modification required. Improving model performance by enhancing data quality frequently yields better results than changing model parameters or adding more data.

Dr. Ng published a comparison of Landing AI's outcomes that illustrates the last point that we just discussed.

As shown in *Figure 1.1*, Landing AI produced three defect detection solutions for their clients. In all three cases, the teams created a baseline model and then tried to improve upon this model using model-centric and data-centric approaches, respectively:

	Steel Defect Detection	Solar Panel Defect Detection	Surface Inspection
Baseline model accuracy	76.2%	75.68%	85.05%
Improvement using model-centric techniques	0% (76.2%)	+0.04% (75.72%)	0% (85.05%)
Improvement using data-centric techniques	+16.9% (93.1%)	+3.06% (78.74%)	+0.4% (85.45%)

Source: A Chat with Andrew on MLOps: From Model-centric to Data-centric AI

Figure 1.1 – Applying data-centric ML – Landing AI's results
(Source: A Chat with Andrew on MLOps: From Model-Centric to Data-Centric AI)

In all three examples, the Landing AI teams were able to achieve the best results by following a data-centric approach over a model-centric approach. In one of three examples, model-centric techniques achieved a tiny 0.04% uplift on the baseline model performance, and in the other two examples, no improvement was achieved.

In contrast, improving data quality consistently led to an improvement in the baseline model, and in two out of three cases quite substantially. The Landing AI teams spent about 2 weeks iteratively improving the training datasets to achieve these results.

Dr. Ng's recommendation is clear: if you want to build relevant and impactful ML models regardless of the size of your dataset, you must put a lot of effort into systematically engineering your input data.

Logically, it makes sense that better data leads to better models and Landing AI's results provide some empirical evidence for the same. Now, let's have a look at why data centricity is the future of ML development.

The components of ML systems

ML systems are comprised of three main parts:

The data-centric approach considers systematic data engineering the key to the next ML breakthroughs for two reasons:

1. Firstly, a model's training data typically carries the most potential for improvement because it is the foundational ingredient in any model.

2. Secondly, the code and infrastructure components of ML systems are much further advanced than our methods and processes for consistently capturing quality data.

Over the last few decades, we have experienced a huge evolution in ML algorithms, data science tools, and compute and storage capacity, and our approach to operationalizing data science solutions has matured through practices such as **ML operations (MLOps)**.

Open source tools such as Python and R make it relatively cheap and accessible for almost anyone with a computer to learn how to produce, tune, and validate ML models. The popularity of these tools is underpinned by the availability of a large number of prebuilt packages that can be installed for free from public libraries. These packages allow users to use common ML algorithms with just a few lines of code.

At the other end of the tooling spectrum, low-code and no-code **automated ML (AutoML)** tools allow non-experts with limited or no coding experience to use ML techniques with a few mouse clicks.

The evolution in cloud computing has provided us with elastic compute and storage capacity that can be scaled up or down relatively easily when demand calls for it (beware of the variable costs!).

In other words, we have solved a lot of the technical constraints surrounding ML models. The biggest opportunity for further upside now lies in improving the availability, accuracy, consistency, completeness, validity, and uniqueness of input data.

Let's take a closer look at why.

Data is the foundational ingredient

Think of the analogous example of a chef wanting to create a world-renowned Michelin Star restaurant. The chef has spent a long time learning how to combine flavors and textures into wonderful recipes that will leave patrons delighted. After many years of practicing and honing their craft, they are ready to open their restaurant. They know what it takes to make their restaurant a success.

At the front of the restaurant, they must have a nicely laid out dining room with comfortable furniture, set up in a way that lets their guests enjoy each other's company. To serve the guests, they need great waiters who will attend to customers' every need, making sure orders are taken, glasses are filled, and tables are kept clean and tidy.

But that's not all. A successful restaurant must also have a fully equipped commercial kitchen capable of producing many meals quickly and consistently, no matter how many orders are put through at the same time. And then, of course, there is the food. The chef has created a wonderful menu full of carefully crafted recipes that will provide their guests with unique and delightful flavor sensations. They are all set to open their soon-to-be award-winning restaurant.

However, on opening night, there is a problem. Mold has gone through some of the vegetables in the pantry and they must be thrown away. Some herbs and spices are out of stock and hard to come by easily. Lastly, the most popular dish on the menu contains red cabbage, but only green cabbage was delivered by the supplier. As a result, the meals are not delightful flavor sensations, but rather bland and average. The chef has built a perfect operation and a wonderful menu but paid too little attention to the most important and hardest-to-control element: the ingredients.

The ingredients are produced outside the restaurant and delivered by several different suppliers. If one or more parts of the supply chain are not delivering, then the final output will suffer, no matter how talented the chef is.

The story of the restaurant illustrates why a more systematic approach to engineering high-quality datasets is the key to better models.

Like the superstar chef needing the best ingredients to make their meals exceptional, data scientists often fall short of building highly impactful models because the input data isn't as good or accessible as it should be. Instead of rotten vegetables, we have mislabeled observations. Instead of out-of-stock ingredients, we have missing values. Instead of the wrong kind of cabbage, we have generic or high-level labels with limited predictive power. Instead of a network of food suppliers, we have a plethora of data sources and technical platforms that are rarely purpose-built for ML.

Part of the reason for this lack of maturity in data collection has to do with the maturity of ML as a capability relative to other disciplines in the computer science sphere. It is common for people with only a superficial understanding of ML to view ML systems the same way they understand traditional software applications.

However, unlike traditional software, ML systems produce variable outputs that depend on a combinatory set of ever-changing data inputs. In ML, the data is part of the code. This is important because the data holds the most potential for varying the final model output. The breadth, depth, and accuracy of input features and observations are foundational to building impactful and reliable models. If the dataset is unrepresentative of the real-world population or scenarios you are trying to predict, then the model is unlikely to be useful.

At the same time, the dataset will determine most of the potential biases of the model; that is, whether the model is more likely to produce results that incorrectly favor one group over another. In short, the input data is the source of the most variability in an ML model and we want to use this variability to our advantage rather than it being a risk or a hindrance.

As we move from data to algorithms and on to system infrastructure, we want the ML system to become increasingly standardized and unvarying. Following a data-centric approach, we want to have lots of the right kind of variability in the data (not noise!) while keeping our ML algorithms and overall operational infrastructure robust and stable. That way, we can iteratively improve model accuracy by improving data quality, while keeping everything else stable.

Figure 1.2 provides an overview of the facets associated with each of the three components of ML systems – data, code, and infrastructure:

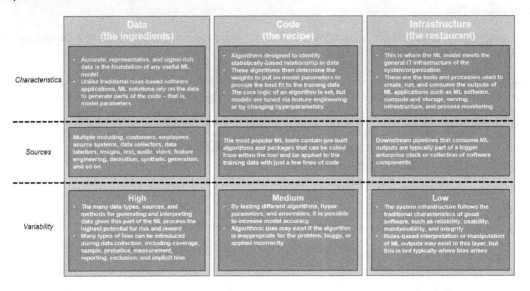

Figure 1.2 – The components of ML systems

Under a data-centric approach, high-quality data is the foundation for robust ML systems. The biggest opportunities to improve an ML model are typically found in the input data rather than the code.

While it makes a lot of sense to focus on data quality over changes to model parameters, data scientists tend to focus on the latter because it is a lot easier to implement in the short term. Multiple models and hyperparameters can typically be tested within a very short timeframe following a traditional model-centric approach, but increasing the signal and reducing the noise in your modeling dataset seems like a complex and time-consuming exercise.

In part, this is because systematically improved data collection typically involves upstream process changes and the participation of various stakeholders in the organization. That is rarely something data scientists can do alone, and it requires the overall organization to appreciate the value and potential of data science to commit the appropriate time and resources to better data collection. Unfortunately, most organizations waste more resources building and implementing suboptimal models based on poor data than the resources it would take to collect better data.

As we will learn in the following sections, a well-designed data-centric approach can overcome this challenge and usually unlocks many new ML opportunities in an organization. This is because data-centric ML requires everyone involved in the data pipeline to think more holistically about the structure and purpose of an organization's data.

To further understand and appreciate the potential of a data-centric approach to model development, let's compare data centricity with the more dominant model-centric approach.

Data-centric versus model-centric ML

So far, we have established that data centricity is about systematically engineering the data used to build ML models. The conventional and more prevalent model-centric approach to ML suggests that optimizing the model itself is the key to better performance.

As illustrated in *Figure 1.3*, the central objective of a model-centric approach is improving the code underlying the model. Under a data-centric approach, the goal is to find a much larger upside in improved data quality:

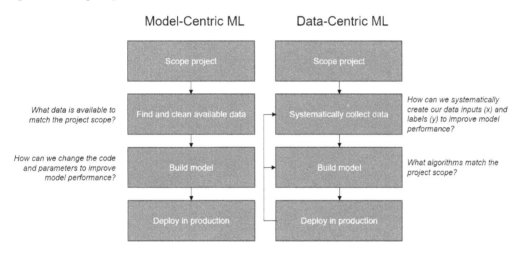

Figure 1.3 – Building ML solutions via model-centric and data-centric workflows

ML model development has traditionally focused on improving model performance mainly by optimizing the code. Under a data-centric approach, the focus shifts to achieving even larger performance enhancements, mainly by iteratively improving data quality. It is important to note that the data-centric approach sits on top of the principles and techniques that underpin model-centric ML, rather than replacing them. Both approaches consider the model and the data critical components of ML solutions. A solution will fail if either of the two is misconfigured, buggy, biased, or applied incorrectly.

Model configuration is an important step under a data-centric approach and in the very short term, it is certainly quicker to seek incremental gains in model performance by optimizing the code. However, as we've discussed, there is limited upside in changing the recipe if you don't have the right ingredients. In other words, the difference between the two approaches lies in where we put our focus and efforts into iteratively improving model performance.

As illustrated in *Figure 1.4*, a model-centric approach treats the data as fixed input and focuses on model selection, parameter tuning, feature engineering, and adding more data as the main ways to improve model performance. A data-centric approach considers the model somewhat static and focuses on improving performance mainly through data quality.

Following a model-centric approach, we attempt to collect as much data as possible to crowd out any outliers in the data and reduce bias – the bigger the dataset, the better. Then, we engineer our model(s) to be as predictive as possible without overfitting.

This is in contrast to a data-centric approach, which has better data collection and labeling at source, on top of model selection and tuning. Data quality is improved even further through outlier detection, programmatic labeling, more systematic feature engineering, and synthetic data creation (these techniques are explained in depth in subsequent chapters):

	Model-centric approach	Data-centric approach
Process	- Collect as much data as possible - Test various modeling techniques - Tune model hyperparameters to find the best fit - Engineer models to handle bias in a subset of the data	- Collect the highest quality data possible - Validate data quality and refine collection - Curate data quality and signal strength to improve model performance across multiple algorithms - Engineer biased subsets to align with the rest of the data
Approach to model improvement	- Add more data to crowd out outliers and capture rare events - Test different algorithms to find best fit on training data - Adjust hyperparameters - Use cross-validation techniques - Use ensemble models or boosting techniques - Engineer new features from existing dataset	- Improve data collection and labeling to ensure data is of appropriate quality and consistency - Use programmatic or algorithmic labeling to populate missing values and identify outliers - Curate existing data to improve signal and reduce noise - Use synthetic data to add rare but impactful observations to the training data
Toolkit	- Feature engineering - Parameter tuning - Subject matter experts to spot outliers - Many algorithms - Big datasets	- Data labelling at source - Programmatic labeling - Systematic and ML-focused data engineering - Subject matter experts and trained curators to label data - Smaller but cleaner datasets

Figure 1.4 – Comparing model-centric and data-centric ML approaches

ML model improvement comes from two areas: improving the code and improving the data. While data collection and engineering processes might sound like a data engineer's job, they really should be a key part of the data scientist's toolbox.

Let's take a look at what's required of data scientists, data engineers, and other stakeholders under a data-centric approach.

Data centricity is a team sport

While it makes a lot of sense to focus on data quality over changes to model parameters, data scientists tend to focus on the latter because it is a lot easier to implement in the short term. Multiple models and hyperparameters can typically be tested within a very short timeframe following a traditional model-centric approach, but increasing the signal and reducing the noise in your modeling dataset seems like a complex and time-consuming exercise that can't easily be dealt with by a small team. Data-centric ML takes a lot more effort across the organization, whereas a model-centric approach largely relies on the data scientist's skills and tools to increase model performance.

Data centricity is a team sport. Data centricity requires data scientists and others involved in ML development to acquire a new set of data quality-specific skills. The most important of these new data-centric skills and techniques is what we will teach you in this book.

Data capture and labeling processes must be designed with data science in mind and performed by professionals with at least a foundational understanding of ML development. Data engineering processes and ETL layers must be structured to identify data quality issues and allow for iterative improvement of ML input data. All of this requires continuous collaboration between data scientists, data collectors, subject matter experts, data engineers, business leaders, and others involved in turning data into insights.

To illustrate this point, *Figure 1.5* compares the data-to-model process for both approaches. Depending on the size and purpose of your organization, there may be a wide range of roles involved in delivering ML solutions, such as data architects, ML engineers, data labelers, analysts, model validators, decision makers, project managers, and product owners.

However, in our simplified diagram in *Figure 1.5*, three types of roles are involved in the process – a data scientist, a data engineer, and a subject matter expert:

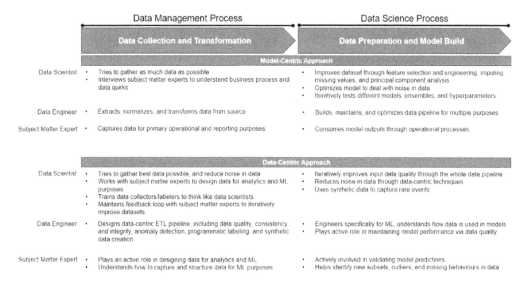

Figure 1.5 – Data-centric versus model-centric roles and responsibilities

Stakeholders at the top of the data pipeline must be active participants in the process for an organization to be good at data collection and engineering for ML purposes. In short, data centricity requires a lot of teamwork.

Under a conventional model-centric approach, data creation typically starts with a data collection process, which may be automated, manual, or a mix of both. Examples include a customer entering details into a web page, a radiographer performing a CT scan, or a call center operator taking a recorded call. At this point, data has been captured for its primary operational purpose, but through the work of the data engineer, this information can also be transformed into an analytical dataset. The typical process requires a data engineer to extract, transform, and normalize the data in a database, data lake, data warehouse, or equivalent.

Once a data scientist gets a hold of the data, it typically goes through several steps to ensure accuracy, consistency, validity, and integrity are maintained. In other words, the data should be ready for use; however, any data scientist knows that this is rarely the case.

A common heuristic in data science is that 80% of the time it takes to build a new ML model is spent on finding, cleaning, and preparing the modeling data for use, while only 20% is spent on analysis and model building. Traditionally, this has been seen as a problem because data scientists are paid to work with the data to build models and perform analyses, and not spend most of their time preparing it.

Following a data-centric approach, data preparation becomes the most important part of the model-building process. Instead of asking *"how might we minimize the time spent on data prep?"*, we instead ask *"how might we systematically optimize data collection and preparation?"* The problem is not that data scientists are spending a lot of time learning and enhancing their datasets. The problem is a lack of connectivity between ML development and other upstream data activities that allow data scientists, engineers, and subject matter experts to co-create faster and more accurate results.

In essence, data centricity is about establishing the processes, tools, and techniques to do this systematically. Subject matter experts are actively involved in key parts of the ML development process, including identifying outliers, validating data labels and model predictions, and developing new features and attributes that should be captured in the data.

Data engineers and data scientists also gain additional responsibilities under a data-centric approach. The data engineer's responsibilities must expand from building and maintaining data pipelines to being more directly involved in developing and maintaining high-quality features and labels for specific ML solutions. In turn, this requires data engineers and data scientists to understand each other's roles and collaborate towards common goals.

In the next section, we will illustrate, through applied examples, the impact a data-centric approach can have on ML opportunities.

The importance of quality data in ML

So far, we have defined what data-centric ML is and how it compares to the conventional model-centric approach. In this section, we will examine what good data looks like in practice.

From a data-centric perspective, good data is as follows[5]:

- **Captured consistently**: Independent (x) and dependent variables (y) are labeled unambiguously

- **Full of signal and free of noise**: Input data covers a wide range of important observations and events in the smallest number of observations possible

- **Designed for the business problem**: Data is designed and collected specifically for solving a business problem with ML, rather than the problem being solved with whatever data is already available

- **Timely and relevant**: Independent and dependent variables provide an accurate representation of current trends (no data or concept drift)

At first glance, this sort of systematic data collection seems both expensive and time-consuming. However, in our experience, highly deliberate data collection is often a foundational requirement for getting the desired results with ML.

To appreciate the importance and potential of data centricity, let's look at some applied examples of how data quality and systematic engineering of features make all the difference.

Identifying high-value legal cases with natural language processing

Our first example of the pivotal importance of data quality comes from an ML solution built by Jonas and Manmohan at a large Australian legal services firm.

ML is a nascent discipline in legal services relative to comparable service industries such as banking, insurance, utilities, and telecommunications. This is due to the nature and complexity of the data available in legal services, as well as the risks and ethics associated with using ML in a legal setting.

Although the legal services industry is incredibly data-rich, data is often collected manually, stored in a textual format, and highly contextual to the particulars of the legal case. This textual data may come in a variety of formats, such as letters from medical professionals, legal contracts, counterparty communications, emails between lawyer and client, case notes, and audio recordings.

On top of that, the legal services industry is a high-stakes environment where a mistake or omission made by one party can win or lose the case altogether. Because of this, legal professionals tend to spend a lot of time and effort reviewing detailed documents and keeping track of key dates and steps in the legal process. The devil is in the detail!

The legal services firm is a no-win-no-fee plaintiff law firm representing people who have been injured or wronged physically or financially. The company fights on behalf of individuals or groups against the more powerful counterparties, such as insurance firms, negligent hospitals or doctors, and misbehaving corporations. The client only pays a fee if they win – otherwise, the firm bears the loss.

In 2022, the business identified an opportunity to use data science to find rare but high-value cases that could then be fast-tracked by specialist lawyers. The earlier in the process that these high-value cases could be identified, the better. So, the goal was to recognize them in the very first interview with prospective clients.

The initial project design followed a conventional model-centric approach. The data science team collected 2 years' worth of case notes from prospective client interviews and created a flag for cases that had later turned out to be high-value (the dependent variable, y). The team also used topic modeling to engineer new features to be included in the final input dataset. **Topic modeling** is an unsupervised ML technique that's used to detect patterns across various documents or text snippets that can be grouped into *topics*. These topics were then used as direct input into the initial model and also as a tool to explain model predictions.

The initial model proved reasonably predictive, but the team faced several challenges that could only be solved by taking a data-centric approach:

- Less than a thousand high-value cases were opened on an annual basis, so this was a *small data* problem, even after oversampling.

- The main predictors were captured from case notes, which were in a semi-structured or unstructured format, and often free text. Although case notes followed some standards, each note taker had used their distinct vocabulary, shortenings, and formatting, making it difficult to create a standardized modeling dataset.

- Because the input data was largely in free-text format, some very important facts were too vague for the model to pick up. For instance, it was important whether the legal case involved more than one injured person as this could change the case strategy altogether. Sometimes, each injured party would be called out explicitly and other times just referred to as *they*.

- Some details were left out of the case notes because they were either assumed knowledge by legal professionals or they would be obvious to a human reading the document as a whole. Unfortunately, this was not helpful to a learning algorithm.

The team decided to take a data-centric approach and formed a cross-functional project team comprising a highly skilled lawyer, a data scientist, a data engineer, an operations manager, and a call center expert. Everyone on the team was an expert in one part of the overall process and together they provided lots of depth and breadth across client experience, legal, data, and operational processes.

Rather than improving model accuracy through feature engineering, the team altered the data capture altogether by designing a set of client questions that were highly predictive of whether a case was high value. The criteria for new questions were as follows:

- It must provide very specific details on whether a case was high value or not

- The format must be easily interpretable by humans and algorithms alike

- It must be easy for the prospective client to answer new questions and the call center operator to capture the information

- It must be easy to create a triaging process around the captured data such that the call center operator can take the right action immediately

The previously mentioned criteria highlight why it is important to involve a wide group of subject matter experts in developing ML solutions. Everyone in the cross-functional team had specific knowledge that contributed to the finer details of the overall solution.

The team identified a handful of key questions that would be highly predictive of whether a case was high-value. These questions needed to be so specific that they could only be answered with a yes, no, or a quantity. For example, rather than looking for the word *they* in a free text field, the call center operator could simply ask *how many people were involved in the incident?* and record only a numeric answer:

Hypothetical legal case notes: Personal Injury Claim - Car Accident

Case notes *before* data-centric improvements

Describe the particulars of the incident:

The client and her daughter were hit from the left by a delivery truck while driving down a main road. Following the incident, the client required immediate medical attention and was transported by ambulance to the nearest hospital. The client was then hospitalized for a duration of three days due to the severity of the injuries sustained.

The client sustained significant injuries in the accident, with two broken femurs. Due to these injuries, the client underwent surgical procedures and necessary treatments during the hospital stay.

Case notes *after* before data-centric improvements

How many people were involved in the incident?

2

Did you go to hospital as a result of the incident?

Yes

How many days where you hospitalized?

3

Did you require ambulance transport?

Yes

Have you been diagnosed with any form of paralysis?

No

Injury type 1

Broken femur

Injury location 1

Right leg

Injury type 2

Broken femur

Injury location 2

Left leg

Figure 1.6 – Hypothetical case notes before and after data-centric improvements

With these questions answered, every prospective case could be grouped into high, medium, and low probability of being a high-value case. The team then built a simple process that allowed call center operators to direct high-probability cases straight into a fast-track process handled by specialized lawyers. Other cases would continue to be monitored using an ML model to detect new facts that may push them into high-value territory.

The final solution was a success because it helped identify high-value cases faster and more accurately, but the benefits of taking a data-centric approach were much broader than that. The focus on improved data collection didn't just create better data for ML purposes. It created a different kind of collaboration between people from across the business, ultimately leading to better-defined processes and a stronger focus on optimizing key moments in the client journey.

Predicting cardiac arrests in emergency calls

Another example comes from an experimental study conducted at the **Emergency Medical Dispatch Center (EMDC)** in Copenhagen, Denmark[6].

A team led by medical researcher *Stig Blomberg* worked to examine whether an ML solution could be used to identify out-of-hospital cardiac arrest by listening to the calls made to the EMDC.

The team trained and tested an ML model using audio recordings of emergency calls generated in 2014, with the primary goal of assisting medical dispatchers in the early detection of cardiac arrest calls.

The study found the ML solution to be faster and more accurate at identifying cases of cardiac arrest as measured by the model's sensitivity. However, the researchers also discovered the following limitations in following a model-centric approach:

- With no ability for structured feedback between ambulance paramedics and dispatchers, there was a lack of *learning* in the system. For instance, it would likely be possible to improve human and machine predictions of cardiac arrest by asking tailored and more structured questions of the caller, such as *"does he look pale?"* or *"can he move?"*.

- Language barriers of non-native speakers impacted model performance. The ML solution worked best with Danish-speaking callers and was worse at identifying cardiac arrests in foreign-accent calls than the human dispatchers who might speak several languages.

- Although the solution had a higher sensitivity (detection of true positives) than human dispatchers, less than one in five alerts were true positives. This created a high risk of alert fatigue among dispatchers, who ultimately bear the risk of acting on ML recommendations or not.

This case study is another prime example of an ML use case that requires a data-centric approach to achieve optimal results while managing risks and ethics appropriately.

Firstly, an ML solution classifying cardiac arrest calls will only ever be based on *small data* due to the nature and complexity of the underlying problem. In this case, it is not necessarily possible to just add more data to improve model performance.

With about 1,000 true cardiac arrests being reported per year from a population of circa 1.8 million people in Greater Copenhagen, even years' worth of call recordings would not add up to a large dataset. Once you consider the many subsets in the data, such as foreign language speakers and those with non-native accents, the data becomes even more fragmented.

The risks and ethical concerns associated with producing wrong predictions (especially false negatives) for life-and-death situations mean that data labels must be carefully curated until any biases are reduced to an acceptable minimum. This requires an iterative process of reviewing data quality and enhancing model features.

Classifying cardiac arrest cases based on a short phone conversation is a complex exercise. It requires subject matter expertise, as well as training and experience from dispatchers and paramedics alike. Building a quality natural language dataset for ML purposes is largely about reducing ambiguity in the interpretation of the signal you're looking for. This, in turn, requires the organization to define what matters in the process that is being modeled by involving subject matter experts in the design. You will learn how this is done in *Chapter 4, Data Labeling is a Collaborative Process*.

Being specific in how questions are asked and answered creates clarity for human agents (in this case, the dispatchers), as well as ML models. This example highlights how data centricity is not just about collecting better data for ML models. It is a golden opportunity to be more deliberate in defining and improving how people work and collaborate across the organization.

The two case studies you have just read through highlight the importance of carefully collecting and curating datasets to be high quality in terms of accuracy, validity, and contextual relevance. In some situations, data quality can be a matter of life and death!

As you will learn in *Chapter 2, From Model-Centric to Data-Centric – ML's Evolution*, there is huge potential for ML to be a fantastic tool in high-stakes domains such as legal services and healthcare, so long as we can manage the risks associated with data quality.

Now that we've discussed the different aspects of data-centric ML, let's summarize what we've learned in this chapter.

Summary

In this chapter, we discussed the fundamentals of data-centric ML and its origins. We also learned how data centricity differs from model centricity, including the roles and responsibilities of key stakeholders in a typical organization using ML. At this point, you should have a solid understanding of data-centric ML and its additional potential compared to a more traditional model-centric approach. Hopefully, this will encourage you to use data-centric ML for your next project.

In the next chapter, we will discover why ML development has been mostly model-centric until now and explore further why data centricity is the key to the next phase of the evolution of AI.

References

1. `https://datacentricai.org/`, viewed 10 July 2022

2. `https://www.andrewng.org/` and `https://www.coursera.org/instructor/andrewng`, viewed 6 July 2022

3. `https://www.youtube.com/watch?v=06-AZXmwHjo`, viewed 2 August 2022

4. `https://ahrefs.com/blog/long-tail-keywords/`, viewed 2 August 2022

5. Derived from *A Chat with Andrew on MLOps – From Model-centric to Data-Centric AI*: `https://www.youtube.com/watch?v=06-AZXmwHjo`, viewed 2 August 2022

6. Zicari et al.: *On assessing trustworthy AI in healthcare: Best practice for machine learning as a supportive tool to recognize cardiac arrest in emergency calls.* Frontiers in Human Dynamics (2021)

2

From Model-Centric to Data-Centric – ML's Evolution

By now, you might be thinking: if data-centricity is essential to the further evolution of AI and ML, how come model-centricity is the dominant approach?

This is a very relevant question to ask, and one we will answer in this chapter. To understand what it takes to shift to a data-centric approach, we must understand the forces that have led to model-centricity being the predominant approach, and how to overcome them.

We will start this chapter by exploring why the evolution of AI and ML has predominately followed a model-centric approach, before diving into the huge opportunity that can be unlocked through data-centricity.

Throughout this chapter, we will challenge the notion that ML requires big datasets and that more data is always better. There is a long tail of *small data* ML use cases that open up when we shift our mindset from *bigger data* to *better data*.

By the end of this chapter, you will have a clear understanding of the progression of ML to date, and know what it takes to build on the current paradigm and achieve even better results with ML.

In this chapter, we will cover the following main topics:

- Exploring why ML development ended up being mostly model-centric
- The opportunity for small-data ML
- Why we need data-centric ML more than ever

Exploring why ML development ended up being mostly model-centric

A short history lesson is in order to truly appreciate why a data-centric approach is the key to unlocking the full potential of ML.

The fields of data science and ML have achieved significant advancements since the earliest attempts to make electronic computers act *intelligently*. The *intelligent* tasks performed by most smartphones today were nearly unimaginable at the turn of the 21st century. Moreover, we are producing more data every single day than was created from the beginning of human civilization to the 21st century – and we're doing so at an estimated growth rate of 23% per annum[1].

Despite these incredible developments in technology and data volumes, some elements of data science are very old. Statistics and data analysis have been in use for centuries and the mathematical components of today's ML models were mostly developed long before the advent of digital computers.

For our purposes, the history of ML and AI starts with the introduction of the first electronic calculation machines during World War II.

The 1940s to 1970s – the early days

Historian and former US Army officer Adrian R. Lewis wrote in his book *The American Culture of War* that "war created the conditions for great advances in technology… without war, men would not traverse oceans in hours, travel in space, or microwave popcorn[2]."

This was indeed the case during World War II, and in the decades that followed. Huge leaps were made in computer science, cryptology, and hardware technology, as fighting nations around the world were racing each other for dominance on every front.

In the 1940s and 1950s, innovations such as compilers, semiconductor transistors, integrated circuits, and computer chips made digital electronic computers capable of performing more complex processes (until this point, a *computer* was predominately the job title of mathematically gifted humans employed to perform complex calculations[3]). This, in turn, led to some early innovations that underpin today's ML models.

In 1943, American scientists Walter Pitts and Warren McCullough created the world's first computational model for neural networks. This formed the basis for other innovations in AI, including Arthur Samuel's self-improving checkers-playing program in 1952 and the **perceptron**, a neural network for classifying images funded by the US Navy and IBM in 1958.

In 1950, British mathematician and computer scientist Alan Turing introduced the *Turing test* for assessing a computer's ability to perform intelligent operations comparable to those of humans. The test was often used as a benchmark for the *intelligence* of a computer and became very influential to the philosophy of AI in general.

The expansion of ML research continued throughout the 1960s, with the development of the nearest neighbor algorithm being one of the most noticeable advances. The work of Stanford researchers Thomas Cover and Peter Hart formed the basis for the rise of the k-nearest neighbor algorithm as a powerful statistical classification method[4].

In 1965, co-founder of Fairchild Semiconductor and Intel, Gordon Moore proposed that processing power and hard drive storage for computers would double every two years, also known as *Moore's law*[5]. Even though Moore's law proved to be reasonably accurate, it would take many decades to reach a point where vast amounts of data could be processed at a reasonable speed and cost.

To put things into perspective, IBM's leading product in 1970 was the System/370 Model 145, which had 500 KB of RAM and 233 MB of hard disk space[6]. The computer took up a whole room and cost $705,775 to $1,783,000[7], circa $5 to $13 million in today's inflation-adjusted dollars. At the time of writing, the latest iPhone 14 has 12,000 times the amount of RAM and up to 2,200 times the amount of hard disk space of the System/370 Model 145, depending on the iPhone configuration[8].

Figure 2.1 – The IBM System/370 Model 145. Everything in this picture is part of the computer's operation (except the clock on the wall). Source: Jean Weber/INRA, DIST

Most of the 1970s are widely recognized as a period of "AI Winter" – a period with very little ground-breaking research or developments in the field of AI. The business world saw little short-term potential in AI, mainly because computer processing power and data storage capacity were underdeveloped and prohibitively expensive.

The 1980s to 1990s – the rise of personal computing and the internet

In 1982, IBM introduced the first personal computer (IBM PC), which sparked a revolution in computer technology at work and in people's homes. It also led to the meteoric rise of companies such as Apple, Microsoft, Hewlett-Packard, Intel, and many other hardware and software enterprises that rode the wave of technological innovation.

The increased ability to digitize processes and information also amplified the corporate world's interest in using stored data for analytical purposes. Relational databases became mainstream, at the expense of network and hierarchical database models[9].

The query language SQL was developed in the 1970s; throughout the 1980s, it became widely accepted as the main database language, achieving the ISO and ANSI certifications in 1986[10].

The explosion in digital information created a need for new techniques to make sense of data from a statistical point of view. Stanford University researchers developed the first software to generate classification and regression trees in 1984, and innovations such as the lexical database WordNet created the early foundations for text analysis and natural language processing.

Personal computers continued to replace typewriters and mainframes into the 1990s, which allowed for the World Wide Web to be formed in 1991. Websites, blogs, internet forums, emails, instant messages, and VoIP calls created yet another explosion in the volume, variety, and velocity of data.

As a result, new methods for organizing more complex and disparate types of data evolved. Gradient boosting algorithms such as AdaBoost and gradient boosting machines were developed by Stanford researchers throughout the late nineties, paving the way for search engines to rank all sorts of information.

The rise of the internet also created a huge business opportunity for those who could organize the information on it. Companies such as Amazon, Alibaba, Yahoo!, and Google were founded during this period to fight for dominance in e-commerce and web search. These companies saw enormous potential in computer science, AI, and ML and invested heavily in developing algorithms to manage their vast stores of information.

The 2000s – the rise of tech giants

ML research picked up pace throughout the 2000s, whether it be in universities or corporate **research and development (R&D)** departments. Computer processing power had finally reached a point where large-scale data processing was feasible for most corporations and researchers.

While internet search engine providers were busy developing algorithms to sort and categorize the ever-growing information being published online, university researchers were creating new tools and techniques that would fuel the evolution of ML.

In 2003, The R Foundation was created to develop and support the open source ML tool and programming language R. As a freely available and open source programming language for statistical computing and graphics[11], R significantly lowered the barrier to entry for researchers looking to use statistical programming in their work and for data enthusiasts wanting to practice and learn ML techniques.

Random Forest algorithms were introduced in 2001 and later patented in 2006 by statisticians and ML pioneers Leo Breiman from the University of California, Berkley, and Adele Cutler from Utah State University[12].

Stanford professor Fei-Fei Li introduced the ImageNet project in 2008 as a free and open image database for training object recognition models[13]. The database was created to provide a high-quality, standardized dataset for object categorization models to be trained and benchmarked on. At the time of writing, ImageNet contains more than 14 million labeled images, organized according to the WordNet hierarchy.

This period also saw the meteoric rise of the network-based business model as a way to create internet dominance. Social media platforms such as LinkedIn, Facebook, Twitter, and YouTube were launched during this period and became supernational tech giants by using ML algorithms to organize information and content created by their users.

As data volumes exploded, so did the need for cheap and flexible data storage. Cloud compute and storage services such as AWS, Dropbox, and Google Drive were launched, while universities joined forces with Google and IBM to establish server farms that could be used for data-intensive research[14]. Increasingly, the availability of processing power was now based on the user's economic justification rather than technical limitations.

2010–now – big data drives AI innovation

Network-based businesses continued to define the direction for the internet and ML development. Search engines, social media platforms, and software and hardware providers invested heavily in R&D activities surrounding AI. As an example, the Google Brain research team was founded in 2011 to provide cutting-edge AI research on big data.

New network-based companies were disrupting industries such as taxis, hotels, travel services, payments, restaurant and food services, media, music, banking, consumer retail, and education – utilizing digital platforms, ML, and vast amounts of consumer data as their powerful competitive advantage.

Traditional research institutions formed tight collaborations with big tech companies, resulting in big leaps in deep learning techniques for audio and image recognition, natural language understanding, anomaly detection, synthetic data generation, and much more.

By 2017, three out of four teams competing in the annual ImageNet Challenge achieved greater than 95% accuracy, proving that image recognition algorithms were now highly advanced.

Powerful algorithms for generating new data were also developed during this golden decade of AI. In 2014, a researcher from the Google Brain team named Ian Goodfellow invented the **Generative Adversarial Network** (**GAN**), a neural network that works by pairing two models against each other[15]. Another form of generative model framework, the **Generative Pre-trained Transformer** (**GPT**), entered the scene in 2018, courtesy of the OpenAI research lab.

With generative models in operation, it was now possible to produce *human-like* outputs such as text snippets, images, artwork, music, and deepfakes – audio and video impersonations of someone's voice and mannerisms.

As *big data*, *ML*, and *AI* became part of the vernacular, the demand for analysts, data scientists, data engineers, and other data professionals increased substantially. In 2011, job listings for data scientists increased by 15,000% year on year[16]. The massive enthusiasm for the potential of data and ML caused analytics pioneers Tom Davenport and DJ Patil to label data science as *the sexiest job of the 21st century* in 2012[17].

Millions of data enthusiasts around the world sought out places to learn the latest ML and data mining techniques. Platforms such as Kaggle and Coursera allowed millions of users to learn through open online courses, enter ML contests, access quality datasets, and share knowledge.

On the tooling front, the proliferation of freely downloadable software programs and packages running on R, Python, or SQL made it relatively easy to access advanced data science techniques at a low cost:

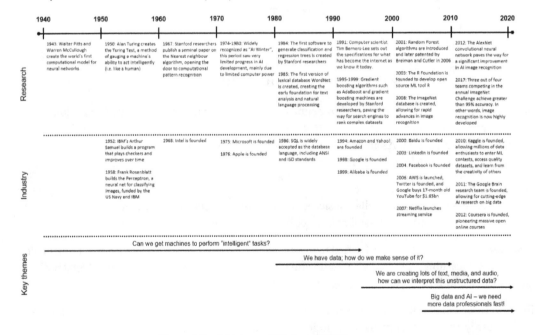

Figure 2.2 – A history of ML from 1940 to now

As the advancements in information technology, data, and AI converged during AI's golden decade of 2010 to 2020, ML model architectures have matured significantly. At this point, most of the opportunities to create better models lie in improving data quality.

Model-centricity was the logical evolutionary outcome

The last eight decades of data science history have followed a logical evolutionary path that has led to model-centricity being the principal approach to ML.

The ideas and mathematical concepts behind ML were imagined long before the technology was mature enough to match them. Before the 1990s, computers were not powerful enough to allow university researchers to evolve the field of ML substantially. These technical limitations also meant that there was limited research conducted for commercial gain by private enterprises during this period.

At the advent of the internet era in the early 1990s, hardware and software solutions were beginning to be advanced enough to eliminate these age-old limitations. The internet also sparked an information revolution that increased the volume and variety of available data enormously. All of a sudden, ML was not just financially viable, it became the driving force behind tech companies such as Amazon, Yahoo!, and Google. With more digital information available than ever before, there was a need to advance the way we interpreted and modeled various kinds of data. In other words, ML research needed a model-centric focus first and foremost.

Throughout the 2000s, a new kind of business model came to dominate our lives. Network-based digital businesses such as social media platforms, search engines, software creators, and online marketplaces created platforms where users could create and interact with content and products. By applying ML to massive amounts of user-generated data, these businesses watched and optimized every interaction along the way.

These "AI-first" big tech businesses were less constrained by data quality or volume. Their constraints lay mostly in fast and affordable compute and storage capacity, and the sophistication of ML techniques. Through in-house research, partnerships with universities, and strategic investments in promising AI technologies, big tech companies have been able to drive the agenda for ML development over the last two decades. What these companies needed primarily was a model-centric approach.

As a result of the model-centric research that has occurred since the mid-1990s, we now have algorithms that can organize all the world's information, identify individuals in a crowd, drive vehicles in open traffic, recognize and generate sound, speech, and imagery, and much more. Our ability to make accurate models *given the input data* is very advanced thanks to this period of innovation.

As data continued to become a more ubiquitous asset, there was a sudden strong need to train more data scientists and other data professionals. Today, there is no shortage of learning opportunities through online learning platforms, university courses, and ML competitions, but they typically have one thing in common: the initial input dataset is predefined.

It makes a lot of sense to teach ML on a fixed dataset. Without a replicable output, it is difficult to verify whether learners have mastered a particular technique, or benchmark different models against each other. However, the natural consequence is that learning is centered around model improvement through model-centric tasks such as model selection, hyperparameter tuning, feature engineering, and other enhancements of the *existing* dataset.

Model-centric skills must be mastered by experienced data scientists, but they are just the foundation of a data-centric paradigm. This is because ML progress comes in four parts:

1. Improving computer power

2. Improving algorithms

3. Improving data

4. Improving measurement

So far, we have made huge progress on items 1 and 2, to a point where they are largely a solved problem for the majority of ML use cases. Most of the opportunity now lies in evolving our approach to improving data and measurement. When we improve our data, we can build better models, but we also unlock the long tail of ML use cases that are often out of reach because we only have a few thousand rows (or less) of data to build our models on.

Unlocking the opportunity for small data ML

The group of tech companies famously labeled *The Big Nine* by author Amy Webb[18] are examples of consumer internet companies that have leveraged big data and AI to build world dominance. Amazon, Apple, Alibaba, Baidu, Meta, Google, IBM, Microsoft, and Tencent dominate in the digital era because they utilize enormous amounts of user data to power their AI systems.

As network-based *AI-first* businesses, they have amassed customers on an unprecedented scale because users are happy to co-create and share their data, so long as it is a net benefit to them. For the Big Nine, getting enough modeling data is rarely a problem, and investing in the most advanced ML capabilities is a virtuous circle that enables more market dominance.

For most other organizations – and ML use cases – this sort of scale is unachievable. As we explored in *Chapter 1, Exploring Data-Centric Machine Learning* the long tail of ML opportunities doesn't offer the option to build models on large volumes of training data because of the following challenges:

*   **The lack of training data observations**: Datasets are smaller in the long tail – typically in the order of only a few thousand rows or less. On top of that, most organizations are capturing data in the non-digitized physical world, which makes it harder to capture and finetune some data points.

- **Dirty data**: Unlike network-based *AI-first* businesses, most organizations generate data through a large variety of sources such as internal (but externally developed) IT systems, third-party platforms, and manual collection by staff or customers. This creates a complex patchwork of data sources that come with a variety of data quality challenges.

- **Risk of bias and unfairness in high-stakes domains**: Poor data quality in high-stakes domains such as healthcare, legal services, education, public safety, and crime prevention may lead to disastrous impacts on individuals or vulnerable populations. For example, predicting whether a person has cancer based on medical images is a high-stakes activity – recommending the next video to watch based on your YouTube history video is not.

- **Model complexity and lack of economies of scale**: Even though there is plenty of value to be found in the long tail, individual ML projects typically need a lot of customization to deal with distinct scenarios. Customization is costly as it creates an accumulation of many models, datasets, and processes that must be maintained pre- and post-model implementation.

- **The need for domain expertise in data and model development**: The combination of small datasets, higher stakes, and more complex scenarios makes it difficult to build ML models without the involvement of subject-matter experts during data collection, labeling and validation, model development, and testing.

It is important to note that many companies have the opportunity to unlock significant value with *small data* ML. For example, only a few organizations will have individual ML projects worth $50 million or more, but many more organizations will have 50 potential ML opportunities worth $1 million each. In practice, this means we must get maximum value out of our raw material if we want smaller projects to become feasible and financially viable.

Dr Andrew Ng, CEO and founder of Landing AI, summarizes these challenges as follows[19]:

> *"In the consumer software Internet, we could train a handful of ML models to serve a billion users. In manufacturing, you might have 10,000 manufacturers building 10,000 custom AI models."*

> *"In many industries, where giant data sets simply don't exist, I think the focus has to shift from big data to good data. Having 50 thoughtfully engineered examples can be sufficient to explain to the neural network what you want it to learn."*

Figure 2.3 illustrates the challenge and opportunity of *small data* ML. While the low-hanging fruits of big data/high-value ML use cases have been picked by *AI-first* businesses, the long tail of small data/moderate value is underexploited. In reality, most ML use cases exist in the long tail of smaller datasets and low economies of scale. A strong focus on data quality is needed to make ML useful when datasets are small:

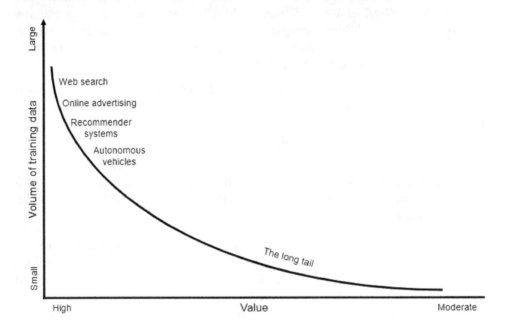

Figure 2.3 – The long tail of ML opportunities

In the next section, we will explore the challenges in working with smaller and more complex datasets, and how you can overcome them.

Why we need data-centric AI more than ever

The leading organizations in AI, such as the Big Nine, have achieved incredible results with ML since the turn of the century, but how is AI being used in the long tail?

A 2020 survey published by MIT Sloan Management Review and Boston Consulting Group concluded that most companies struggle to turn their vision for AI into reality. In a survey of over 3,000 business leaders from 29 industries in 112 countries, 70% of respondents understood how AI can generate business value and 57% had piloted or productionized AI solutions. However, only 1 in 10 had been able to generate significant financial benefits with AI.[20]

The survey authors found that companies that were realizing significant financial benefits with AI had built their success on two pillars:

- They had a solid foundation of the right data, technology, and talent.
- They had defined several effective ways for humans and AI to work and learn together. In other words, they had created an iterative feedback loop between humans and AI, going from data collection and curation to solution deployment.

Why are these two pillars critical to success with ML and AI? Because the ML model is only a small part of an ML system.

In 2015, Google researchers Sculley et al.[21] published a seminal paper called *Hidden Technical Debt in Machine Learning Systems*, in which they describe how *"only a small fraction of real-world ML systems are composed of the ML code… the required surrounding infrastructure is vast and complex."*

In traditional information technology jargon, *technical debt* refers to the long-term costs incurred by cutting corners in the software development life cycle. It's the hardcoded logic, the missing documentation, the lack of integration with other platforms, inefficient code, and anything else that is a roadblock to better system performance and future improvements. Technical debt can be "paid down" by removing these issues.

ML systems are different in that they can carry technical debt in code, but they also have the added complexity that technical debt may exist in the data components of the system. Input data is the foundational ingredient in the system and the data is variable. Because ML models are driven by weighted impacts from many features in both data and code, a change in one variable may change the logical structure of the rest of the model. This is also known as the CACE principle: *Changing Anything Changes Everything*.

As illustrated in *Figure 2.4*, a productionized ML system is much more than the model code. In a typical ML project, it is estimated that only 5-10% of the overall system is the model code[22]. The remaining 90-95% of the solution is related to data and infrastructure:

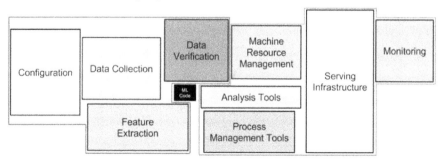

Figure 2.4 – ML systems are much more than code. Source: Adapted from Sculley et al., 2015

As Sculley et al. described, the data collection and curation activities in an ML solution are often significantly more resource-intensive than direct model development activities. Given this, data engineering should be a data scientist's best friend. Yet, there is a disconnect between the importance of data quality and how most ML solutions are developed in practice.

The cascading effects of data quality

In 2021, Google researchers Sambasivan et al.[23] conducted a research study of the practices of 53 ML practitioners from the US, India, and East and West Africa working in a variety of industries. The study participants were selected from high-stakes domains such as healthcare, agriculture, finance, public safety, environmental conversation, and education.

The purpose of the study was to identify and describe the downstream impact of data quality on ML systems and present empirical evidence of what they call *data cascades* – compounding negative effects stemming from data quality issues.

Data cascades are caused by conventional model-centric ML practices that undervalue data quality and typically lead to invisible and delayed impacts on model performance – in other words, ML-specific technical debt. According to the researchers, data cascades are highly prevalent, with 92% of ML practitioners in the study experiencing one or more data cascades in a given project.

The causes of data cascades fit into four categories explained in the following subsections.

The perceived low value of data work and lack of reward systems

There are often two underlying reasons for the lack of available data in the long tail of ML opportunities:

- Firstly, the events being modeled are bespoke and rare, so there is a physical limit to the amount of data that can be collected for a given use case
- Secondly, data collection and curation activities are considered relatively expensive and difficult, especially when they involve manual collection

In truth, most data-related work is not done by data scientists. The roles directly responsible for creating, collecting, and curating data are often performing these tasks as a secondary duty in their job. The responsibility of collecting high-quality data is frequently at odds with other duties because of competing priorities, time constraints, technical limitations of collection systems, or simply a lack of understanding of how to carry out good data collection.

Take, for example, a hospital nurse who is responsible for a wide variety of tasks relating to the care of patients, some of which are data collection. High-quality data in healthcare has the potential to create huge benefits for patients and healthcare providers around the world if it can be aggregated and generalized through ML. However, for the individual nurse, there is more incentive to do the minimum required to document patient status and medical interventions, so more time can be spent on primary patient care. The typical result of this kind of scenario is suboptimal data collection in terms of depth of detail and consistency of labeling.

ML practitioners face a similar challenge further downstream. Sambasivan et al. describe how business and project goals such as cost, revenue, time to market, and competitive pressures lead data scientists to hurry through model development, leaving insufficient room for data quality and ethics concerns. As one practitioner states, *everyone wants to do the model work, not the data work.*

Lack of cross-functional collaboration

When it comes to high-stakes or bespoke ML projects, subject-matter experts are often critical participants in upstream data collection as well as the ultimate consumers of model outputs.

On the face of it, subject-matter experts should be very willing to participate actively in ML projects because they get to reap the benefits of useful models. However, the opposite is often the case.

A requirement to collect additional information for ML purposes typically means that data collectors and curators have to work harder to get their job done. It can be difficult for frontline workers with limited data literacy to appreciate the importance of data collection, and unfortunately, the cascading effect of this conduct shows up much later in the project life cycle – often after deployment.

Data scientists should also play a critical role in data collection as they will make many decisions on how to interpret and manipulate datasets during model development. Therefore, an ML practitioner's curiosity and willingness to understand the technical and social contexts of a given domain is a critical part of any project's success. It is the invisible glue that makes ML solutions relevant and accurate.

Unfortunately, data scientists often lack domain-specific expertise and rely on subject-matter experts to validate their interpretation of datasets. If ML practitioners do not constantly question their assumptions, rely too heavily on their technical expertise, and take the accuracy of input data for granted, they will miss the finer points of the context they're trying to model. When this happens, ML projects will suffer from data cascades.

Insufficient cross-functional collaboration results in costly project challenges such as additional data collection, misinterpretation of results, and lack of trust in ML as a relevant solution to a given problem.

Educational and knowledge gaps for ML practitioners

Even the most technically skilled ML practitioners may fail to build useful models for real-life scenarios if they lack end-to-end knowledge of ML pipelines. Unfortunately, most learning paths for data scientists lack appropriate attention to data engineering practices.

Graduate programs and online training courses are built on clean datasets, but real life is full of dirty data. Data scientists are simply not trained in building ML solutions from scratch, including data collection design, data management, and data governance processes, training data collectors, cleaning dirty data, and building domain knowledge.

As a result, data engineering and MLOps practices are poorly understood and under-appreciated by those who are directly responsible for turning raw data into useful insights.

Lack of measurement of and accountability for data quality

Conventional ML practices rely on statistical accuracy tests, such as *precision* and *recall*, as proxies for model *and* data quality. These measures don't provide any direct information on the quality of a dataset as it pertains to representing specific events and relevant situational context. The lack of standardized approaches for identifying and rectifying data quality issues early in the process makes data improvement work reactive, as opposed to planned and aligned to project goals.

The much-used management phrase *what gets measured gets managed* is also true in a data quality setting. Without appropriate processes in place for identifying data quality issues, it is difficult to incentivize and assign accountability to individuals for good data collection.

The importance of assigning accountability for data quality in high-stakes domains is underpinned by the fact that model accuracy typically has to be very high, based on small datasets. For example, a poorly performing model in a low-risk and data-rich industry, such as online retailing or digital advertising, can be modified relatively quickly given the automated and persistent nature of data collection.

ML models deployed in the long tail are often harder to validate because of a much lower frequency of events. At the same time, high-stakes domains typically demand a higher model accuracy threshold. Online advertisers can probably live with an accuracy score of 75%, but a model built for cancer diagnosis typically has to have an error rate of less than 1% to be viable.

Avoiding data cascades and technical debt

The pervasiveness of data cascades highlights a larger underlying problem: the dominant conventions in ML development are drawn from the practices of *big data* companies. These practices have been developed in an environment of plentiful and expendable data where each user has one account[24]. Combine this with a culture of *move fast and break things*[25] while viewing data work as undesirable drudgery, and you have an approach that will fail in most high-stakes domains.

The cascading effects of poor data are opaque and hard to track in any standardized way, even though they occur frequently and persistently. Fortunately, data cascades are also fixable. Sambasivan et al. define the concept of *data excellence* as the solution: a cultural shift toward recognizing data management as a core business discipline and establishing the right processes and incentives for those who are a part of the ML pipeline.

As data professionals, it's up to us to decide whether ML should remain a tool for the few or whether it's time to allow projects with smaller financial value or higher stakes to become viable. To do this, we must strive for data excellence.

Now, let's summarize the key takeaways from this chapter.

Summary

In this chapter, we reviewed the history of ML to give us a clear understanding of why model-centric ML is the dominant approach today. We also learned how a model-centric approach limits us from unlocking the potential value tied up in the long tale of ML opportunities.

By now, you should have a strong appreciation for why data-centricity is needed for the discipline of ML to achieve its full potential but also recognize that it will require substantial effort to make the shift. To become an effective data-centric ML practitioner, old habits must be broken and new ones formed.

Now, it's time to start exploring the tools and techniques to make that shift. In the next chapter, we will discuss the principles of data-centric ML and the techniques and approaches associated with each principle.

References

1. `https://www.idc.com/getdoc.jsp?containerId=prUS47560321`, viewed on 23 September 2022

2. Lewis, A. R., 2006, *The American Culture of War*, Routledge, New York, USA

3. `https://www.nasa.gov/feature/when-the-computer-wore-a-skirt-langley-s-computers-1935-1970`, viewed on 23 September 2022

4. `https://www.historyofdatascience.com/k-nearest-neighbors-algorithm-classification-and-regression-star/`, viewed on 23 September 2022

5. `http://large.stanford.edu/courses/2012/ph250/lee1/docs/Excepts_A_Conversation_with_Gordon_Moore.pdf`, viewed on 23 September 2022

6. `https://www.businessinsider.com/ibm-1970-mainframe-specs-are-ridiculous-today-2014-5`, viewed on 22 September 2022

7. `https://www.ibm.com/ibm/history/exhibits/mainframe/mainframe_PP3145.html`, viewed on 22 September 2022

8. `https://www.apple.com/au/iphone-14/specs/`, viewed on 23 September 2022

9. `https://www.quickbase.com/articles/timeline-of-database-history`, viewed on 24 September 2022

10. `https://www.dataversity.net/brief-history-database-management/`, viewed on 24 September 2022

11. `https://www.r-project.org/about.html`, viewed on 24 September 2022

12. `https://www.historyofdatascience.com/leo-breiman-statistics-at-the-service-of-others/`, viewed on 24 September 2022

13. `https://www.image-net.org/about.php`, viewed on 24 September 2022

14. `https://www.dataversity.net/brief-history-cloud-computing/`, viewed on 25 September 2022

15. `https://thenextweb.com/news/2010-2019-the-rise-of-deep-learning`, viewed on 25 September 2022

16. `https://www.dataversity.net/brief-history-data-science/`, viewed on 25 September 2022

17. `https://hbr.org/2012/10/data-scientist-the-sexiest-job-of-the-21st-century`, viewed on 25 September 2022

18. Webb, A., 2019, *The Big Nine: How Tech Titans and Their Thinking Machines Could Warp Humanity*, Hachette Book Group, New York, USA

19. `https://spectrum.ieee.org/andrew-ng-data-centric-ai`, viewed on 25 September 2022

20. Ransbotham, S., Khodabandeh, S., Kiron, D., Candelon, F., Chu, M., and LaFountain, B., *Expanding AI's Impact With Organizational Learning*, MIT Sloan Management Review and Boston Consulting Group, October 2020

21. `https://papers.nips.cc/paper/2015/file/86df7dcfd896fcaf2674f757a2463eba-Paper.pdf`, Sculley et al., 2015, viewed 23 July 2022,

22. Yang, K., 2022, *Landing AI – Moving Beyond the Software Industry*, `https://community.ai-infrastructure.org/public/videos/landing-ai-ai-moving-beyond-the-software-industry-2022-09-30`

23. Sambasivan, N., Kapania, S., Highfill, H., Akrong, D., Paritosh, P., Aroyo, L., 2021, *Everyone wants to do the model work, not the data work: Data Cascades in High-Stakes AI*

24. `https://hbr.org/2019/01/the-era-of-move-fast-and-break-things-is-over`, viewed on 8 October 2022

Part 2: The Building Blocks of Data-Centric ML

In this part, we lay the groundwork for data-centric ML with four key principles that underpin this approach, giving you essential context before exploring specific techniques. Then we explore human-centric and non-technical approaches to data quality, examining how expert knowledge, trained labelers, and clear instructions can enhance your ML output.

This part has the following chapters:

- *Chapter 3, Principles of Data-Centric ML*
- *Chapter 4, Data Labeling Is a Collaborative Process*

3
Principles of Data-Centric ML

In this chapter, you will learn the key principles of data-centric ML. We'll cover the foundational principles of data-centricity in this chapter to provide a high-level structure and framework to work through and refer to throughout the rest of this book. These principles will give you important context – or the *why* – before we dive into the specific techniques and approaches associated with each principle in the following chapters – or the *what*.

As you read through the principles, remember that data-centric ML is an extension – and not a replacement – of a model-centric approach. Essentially, model-centric and data-centric techniques work together to glean the most value from your efforts.

By the end of this chapter, you will have a good understanding of each of the principles and how they work together to form a framework for data-centricity.

In this chapter, we'll cover the following topics:

- Principle 1 – data should be the center of ML development
- Principle 2 – leverage annotators and **subject-matter experts** (**SMEs**) effectively
- Principle 3 – use ML to improve your data
- Principle 4 – follow ethical, responsible, and well-governed ML practices

Sometimes, all you need is the right data

A few years ago, I (Jonas) was leading a team of data scientists tasked with an interesting but challenging problem. The financial services business we worked for attracted many new online visitors wanting to open new accounts with us through the company's website. However, a significant number of potential customers couldn't complete the account opening process for unknown reasons, which is why the company turned to its data scientists for help.

This problem of unopened accounts and lost customers was multifaceted, but we were determined to find every needle in the haystack. The account opening process was rather straightforward, designed to make it easy for someone to open a new account in less than 10 minutes with no support. For the customer, the steps were as follows:

1. Enter personal details.
2. Verify identity.
3. Verify contact details.
4. Accept the terms and conditions and open an account.

This process worked most of the time, but things were going wrong in *steps 2* and *3* for a significant proportion of applicants. If someone's identity couldn't be verified online (*step 2*), the individual would have to be verified in person, which was an obvious detractor for many, and it caused a significant drop-off.

The problems arising in *step 3* were less obvious. About 10% of users would quit their journey at this point, even though most of the hard work had already been done. Why would someone go through this whole process and then decide not to proceed after all?

We collected all the relevant data points we could get our hands on, but unfortunately, we didn't have a very deep dataset to work on because the account opening process was so simple and these were new customers. We profiled our dataset and used various supervised and unsupervised ML techniques to tease out any behaviors that correlated with accounts not opening, but nothing stuck out in our analysis.

We decided to dig deeper. Since these clients shared their contact information, we could match their phone numbers with our phone call records and obtain the recorded conversations with matching phone numbers. We pulled out hundreds of call recordings and started listening in.

Soon after, a clear pattern emerged: "I clicked the **Verify contact details** button, but never received a verification code," said one recorded caller. "I've waited for 10 minutes, but the code hasn't come through yet," said another. Users weren't getting through because they weren't sent the final verification code as a text message – even when it was resent by call center agents. But this wasn't the case for all new users, so what was going wrong for this particular group?

As we continued to listen to call recordings, another faint signal emerged: "I shouldn't have come back," said one user. "Your systems haven't gotten any better since the last time I was here," said another.

We had a look at closed customer accounts and sure enough, these people had been customers of ours in the past. The issue was simply that the enterprise system was treating these users as existing customers and therefore not sending out the required text messages, no matter how many times it was prompted by users or staff. The issue was occurring around 200 times a week, meaning the business was missing out on 10,000 new customers a year. Why didn't anyone pick up on this issue earlier?

Only a proportion of the 200 occurrences would generate a call, and with hundreds of call center staff on duty throughout the week, it seemed like a rare glitch that only happened now and then. No one individual could see the issue because it was too infrequent and impossible for our models to flag. After all, the initial dataset had too much noise and not enough signal.

We only got to the bottom of it and found our needle in the haystack because we followed the four principles of data-centricity discussed in this chapter. Let's explore each of these principles in more detail.

Principle 1 – data should be the center of ML development

As we discussed in *Chapter 2, From Model-Centric to Data-Centric – ML's Evolution,* the predominant model-centric approach is lacking in several ways: computing and storage have been commoditized, algorithms have become practically automated and highly data-dependent, models are accessible but less malleable, and deep learning and AutoML tools are available everywhere. But the data? Well, that's still the wildcard.

Rather than relying on powerful computing and storage environments and sophisticated algorithms that demand excess amounts of data to give us the incremental uplift in model accuracy, a better approach is to be driven by data – specifically, by the data that is available and relevant to the problem at hand.

Data is unique to every company, problem, and situation, and the data-centric paradigm recognizes this by putting the spotlight and development efforts on the data before the model. Data is no longer a static asset that can be collected at the beginning and forgotten about; it is now a unique commodity that needs to be leveraged to its full potential to make better predictions. We will argue that in many cases, a company's proprietary data is its only truly unique competitive advantage – so long as it's leveraged.

By focusing on the data, data-centricity helps companies distinguish themselves from their competitors. Most companies have access to the same algorithms and plenty of computing and storage, but the data they use and the insights they gain from that data can give them a decisive edge over the competition.

In our observation, focusing on data quality also brings substantial benefits to an organization besides getting better data. Focusing on data quality means going beyond simple data refinement because quality data is a critical component of business operations.

For data quality to improve, there is typically a need to digitize and automate processes and to create strong accountability for process adherence. Strong data governance processes will assign ownership, stewardship, and accountability for data quality, which, in turn, relies on data collection standards and processes to be followed, measured, and managed.

Data quality is often a symptom of the quality of underlying processes and adherence to these processes. If an organization is good at collecting high-quality data, it is also likely to have good processes more generally. As companies improve data collection, they drive better accountability, accuracy, reliability, and overall consistency in their operations. Hence, focusing on data quality can have far-reaching consequences beyond improved data integrity and reliability. It is an essential driver of operational excellence.

If we think back to the example of missing verification codes discussed at the beginning of this chapter, no amount of model selection, parameter tuning, or feature engineering on the existing dataset would have revealed the root cause of the problem.

This issue could only be discovered and solved by collecting the right data for the problem at hand. In this case, the missing data points were as follows:

- Verification codes weren't being received
- This was only the case for previous customers who returned to open a new account

The discovery of the verification code issue resulted in two key changes to the way the business operated. Firstly, the IT department fixed the code responsible for triggering verification codes being sent, which, all else being equal, resulted in a substantial uplift in new account openings. Secondly, the call center team established a central process for logging client tech issues, no matter how small, so we could discover *the tip of the iceberg* of any new system issues. In other words, it was now accepted culture that collecting high-quality data is central to improving operational processes.

This was a mindset shift for frontline staff in two ways. Firstly, there was a newfound appreciation for data as a powerful asset that could be aggregated and analyzed to understand the bigger picture of their work. Secondly, frontline teams now felt empowered: if I do my bit to capture and call out important issues, there is a chance we can fix them.

Our data scientists also gained a different appreciation for data collection and curation as a key part of *their* role. Seeing the impact they could have by walking in the shoes of customers and frontline staff created a profound shift in the way the team solved problems. Rather than accepting data (quality) as given, data engineering now permeated all steps of the model development and deployment process.

A checklist for data-centricity

To stay true to the first principle of data-centric ML, it is incredibly valuable to have a checklist of data-focused tasks to complete as you work through an ML project. Here is our checklist, spread across the five steps in the model development life cycle.

Step 1 – identify the business problem, scope the project, and define the data needs

The first part of any ML project should always be to clearly define the problem you're trying to solve; this should be done in collaboration with key stakeholders such as end users and SMEs. Identifying data gaps will be a lot easier when you have a clear definition of the problem you're solving, and what success looks like when the problem has been solved.

A strong modeling dataset contains both *content* and *context*. Content is a specific object, event, or status you're measuring and context describes the circumstances in which the object, event, or status occurred. In our previous missing verification code example, the content is the *(missing) verification code* and the context is *for former, returning customers.*

A critical element in defining the project scope is outlining the process you're trying to model. We do this by getting relevant end users and SMEs in a room with data scientists for as long as it takes to map out the process or situation underlying the business problem. This allows all participants to get a deep understanding of the content and context of the problem, while also identifying important data points needed for the model build.

Here is a checklist of questions to consider during this step:

- Have we clearly defined the problem we're trying to solve?

- Is it the right problem to solve in the first place?

- What outcomes are we looking to achieve by solving the problem?

- Have we mapped out the key parts of the problem with SMEs?

- What are the critical steps or moments in the process according to SMEs, and does our data capture these appropriately?

- Do our data points contain both *content* and *context*?

- What biases could arise from the solution that we need to look out for later?

- Will any groups or segments be treated unfairly as a result of these biases? How can we identify these during the validation phase (*step 4*)?

- Is it legal and ethical to use all features in our dataset?

- Will the outputs be internally or externally audited?

Step 2 – prepare and label data

For many data scientists, data preparation is a dreaded task. We certainly agree that data preparation can be both repetitive and time-consuming, but as proponents of data-centric ML, we encourage you to embrace it as the most important part of the job.

Data preparation involves collecting, cleaning, structuring, enhancing, and augmenting your input data to increase the signal and reduce noise in the dataset. These tasks can be both technically challenging and rewarding – especially when you start seeing those AUC scores increase. By now, you are aware that this part of the process is likely to give you very powerful modeling outcomes if you put in the right kind of effort.

The following checklist is useful for guiding you through the data preparation process. We will teach you how to do these tasks in plenty of detail throughout the rest of this book.

Here are the checklist questions:

- Have we performed a technical validation of data quality? (See *Chapter 5, Techniques for Data Cleaning.*)

- Can we enhance the strength of our dataset by cleaning it? (See *Chapter 5, Techniques for Data Cleaning.*)

- Do we need to collect additional data or enhance the quality of the existing dataset using human labelers? (See *Chapter 4, Data Labeling Is a Collaborative Process.*)

- Do we need to define specific labeling rules and train SMEs? (See *Chapter 4, Data Labeling Is a Collaborative Process.*)

- Can we improve data quality or impute missing values using programmatic labeling? (See *Chapter 6, Techniques for Programmatic Labeling in Machine Learning.*)

- Should we use synthetic data to augment or enhance certain classes in the data? (See *Chapter 7, Using Synthetic Data in Data-Centric Machine Learning.*)

- Do we need to preserve the privacy of individuals in the dataset? (See *Chapter 7, Using Synthetic Data in Data-Centric Machine Learning.*)

- Does our dataset contain biased classes and do we need to adjust these? (See *Chapter 8, Techniques for Identifying and Removing Bias.*)

- Does our dataset contain enough of the right kinds of rare events? Do we need to add more or remove outliers? (See *Chapter 9, Dealing with Edge Cases and Rare Events in Machine Learning.*)

- Can we engineer new features from the existing dataset?

Step 3 – train the model

The model training phase is where data-centric and model-centric ML principles come together to create synergy. Again, it is important to highlight that you should not discard everything you already know about how to build and enhance ML models based on a model-centric approach. Data-centricity simply gives you an additional set of tools in your toolbox and allows you to amplify the impact of your models.

Feature selection is an important part of this synergistic process because it filters out features that aren't useful (and in the worst case, problematic) for your model. Generally speaking, it is desirable to have fewer attributes contributing to your model because it reduces unwanted noise and makes the model easier to explain.

It is important to consider feature selection as part of the model selection process because a model and its input data go hand in hand to produce predictions. They are intrinsically linked to each other. Practically speaking, this means you should pick the features *with* the model and not use a static dataset of pre-selected features to choose between models.

Here are the checklist questions:

- Do you suspect your data to be dirty (for example, wrong labels, missing values, meaningless patterns, or irrelevant outputs)?

- Can we improve model accuracy by improving the quality of the dataset? The long list of data-centric techniques outlined in this book is designed to help you with this task!

- Do our engineered features suggest any relationships in the data that require us to collect additional data (features or observations)?

- Do our (engineered) features suggest any relationships in the data that we should verify with SMEs? Rather than assuming our new features are correct, any influential correlations should be cross-checked with SMEs to ensure their relevance and validity within the context of the solution.

- Can we reduce the number of features in our model without losing predictive power? By applying dimensionality reduction techniques or feature selection methods, we may be able to decrease the number of features in our model without significantly compromising its predictive accuracy.

Step 4 – evaluate performance, fairness, and bias

The data-centric approach to ML puts a strong emphasis on detecting bias and fairness issues during model evaluation. To validate the accuracy of an ML model, you should still start with traditional validation tasks such as splitting data into training and testing sets, performing cross-validation, and producing confusion matrices. The following checklist assumes that you will already be doing these tasks, using standard performance evaluation using metrics such as accuracy, precision, recall, F1 score, and more.

Bias detection is an important tool for uncovering potential unfairness and discrimination in ML models. This can be done by creating confusion matrices for each subgroup separately to compare false positive and false negative rates across groups, and assessing demographic parity (equal representation) and equal opportunity (equal true positive rate) or equal odds (equal false positive rate) across groups. Disparities between subgroups, such as those related to gender or race, are common examples of sources of bias or unfairness.

Here are the checklist questions:

- Can we improve model performance by improving data quality or collecting new features?

- Can we detect any bias or unfairness toward particular groups or segments?

- Are there any large correlations between sensitive attributes and predictions made by the model?

- How does the model perform on unseen data concerning fairness and bias?

We'll cover techniques for identifying and removing bias in *Chapter 8, Techniques for Identifying and Removing Bias*.

Step 5 – deploy and monitor

Effective monitoring of ML models also relies on data-centric principles. When monitoring model performance, it is important to include data quality, data coverage, data relevance, and labeling consistency metrics.

Data quality refers to the accuracy, completeness, and consistency of data, while data coverage refers to having enough data points to make confident predictions in the first place. Data relevance ensures that the data used to train the model is suitable for the task. Finally, data labeling consistency ensures that the data points used for training the model have correct labels.

Several techniques and tools can help data scientists monitor ML models effectively. For example, data drift detection helps detect changes in data characteristics, such as mean, variance, and distribution. Similarly, outlier detection helps identify data points that differ significantly from the common distribution. Also, bias detection techniques help identify and correct instances of bias in ML models.

In addition to relying on reporting and metrics to monitor ML models, it is crucial to understand that monitoring is an ongoing process that requires the involvement of stakeholders beyond the data science team. Stakeholders may include SMEs, business owners, and end users. These stakeholders should collaborate to evaluate the model's performance, interpret the results, and identify any issues that need to be addressed.

Here are the checklist questions:

- Are the data sources used in the model automated, consistent, and reliable?
- Have we designed a monitoring and reporting plan that captures failures, biases, and drift?
- Does our monitoring quantify data quality, data coverage, data relevance, and labeling consistency?
- Have we set up a mechanism for end users to provide continuous feedback on model performance?

Our checklist questions are designed to make you think about data quality, and the impacts thereof, at each step in the model development process. In other words, they are an addition to – and not a replacement for – more model-centric development tasks.

Data-centricity requires a mindset shift from "I'll build the best model with this data" to "How can we make the best dataset to solve this particular problem?" To do that, we need the whole organization involved in a coordinated effort. This brings us to the second principle of data-centric ML.

Principle 2 – leverage annotators and SMEs effectively

No matter where we are in the AI hype cycle when you read this, it is unlikely that AI and ML development has evolved past the point where human input and labeling are needed.

In recent years, we have experienced a large increase in the sophistication of AI technologies, especially in the field of generative AI. Despite this, it remains a fact that even the most powerful and revolutionary AI technologies, such as ChatGPT, rely on small armies of human labelers to refine and advance their capabilities.

These individuals review and annotate data samples, which are then fed back into the model to improve its understanding of natural language and context. Some of the key methodologies and techniques that are employed by human labelers include the following:

- **Domain expertise**: Labelers with subject matter expertise can provide valuable insights and annotations that help the model better comprehend specific topics and domains.

- **Active learning**: This approach involves prioritizing data samples that the model finds ambiguous or challenging, enabling labelers to focus on areas where their input can have the greatest impact.

- **Diversity of perspectives**: By involving labelers from diverse backgrounds and with varied experiences, the model can be exposed to a broader range of linguistic nuances, cultural contexts, and perspectives, improving its overall performance.

- **Quality control**: Regular audits and evaluations of labeler output can help ensure consistent annotation quality and adherence to guidelines, which is essential for effective model training.

In short, human annotators are integral to ML, and the quality of our models depends on our ability to train, organize, and collaborate with these annotators. Broadly speaking, there are three ways to leverage SMEs in the ML development process:

- As direct labelers of data points

- As verifiers of output quality and detectors of undesired outputs such as toxic content

- As knowledge experts who can help us codify labeling rules

Leveraging SMEs effectively requires a mindset shift from just creating labeling rules for annotators (although this is still important) to using the combined strengths of SMEs and data scientists to cover the problem space through well-defined labeling rules.

Our experience is that following this approach gives us much more than robust labeling functions. It helps us track down ambiguous examples and sharpen model performance, but just as importantly, it allows data scientists and SMEs to collaborate. As data scientists learn about the subject matter and SMEs learn how data science works, it creates a flywheel effect leading to new ideas, insights, and knowledge.

Let's explore each of the three human labeling approaches in more detail.

Direct labeling with human annotators

The primary benefit of human-annotated data is its accuracy. Humans can recognize patterns and subjectivity in ways that computers cannot. This means that the labels assigned to the data may be more accurate than those generated by automated processes. Additionally, humans can provide context to the data that would otherwise be lost in an automated process.

Human-annotated data also offers greater flexibility than automated processes. Annotators can customize their labeling process according to specific needs or requirements, allowing them to tailor the annotations to fit their project's goals. This makes it easier for machines to interpret the data accurately and quickly.

Finally, human-annotated data can be cost-effective compared to other methods of labeling. This is mainly true when datasets are small- to medium-sized.

Small datasets might consist of a few hundred to a few thousand observations, often manageable by a small team of human annotators. Medium-sized datasets may have tens of thousands of observations. While still possible to manually label, the complexity and time required start to increase, making it less economically viable.

When faced with larger datasets, manual annotation can become repetitive and prone to mistakes due to the sheer volume and potential complexity of the data. At this scale, the intricacies in the data could also increase, requiring a more nuanced understanding that may be challenging for human annotators to maintain consistently.

For larger or more complex datasets, we recommend going down the path of programmatic labeling, which we'll discuss next. Interestingly, a hybrid approach can also be effective, where a subset of the large dataset is manually labeled to serve as training data for the programmatic labeling algorithm. This way, you can leverage the accuracy of human annotation and the scalability of ML, ensuring high-quality labels, even for large datasets.

Think back to the story of the missing verification codes we outlined at the beginning of this chapter. Once we had isolated the phone calls that were related to the yet-to-be-discovered issue, we chose to manually listen to hundreds of calls rather than use ML techniques to pick up themes. Why?

Because we wanted to make sure we understood the *content* and the *context* of these interactions and a human was just more likely to do that job well. At the same time, we were only listening to a few hundred calls, not millions, so human annotation was the most cost-effective way to find the signal in the noise and pinpoint our needle-in-the-haystack issue.

While labeling by humans is an essential part of the data-centric approach, there are several pitfalls and mistakes to avoid when using human labelers. In *Chapter 4, Data Labeling Is a Collaborative Process*, we will teach you how to get the most out of SMEs and human annotators while managing the potential downsides.

Verifying output quality with human annotators

As mentioned previously, even very sophisticated AI solutions such as ChatGPT are heavily reliant on human annotators to guide algorithms to the optimal outcome. ChatGPT has been built on a mix of supervised learning and a technique called **reinforcement learning from human feedback (RLHF)**.

Reinforcement learning is an area of ML where an agent learns to make decisions by interacting with an environment. The agent's objective is to select actions that maximize the cumulative reward over time. However, defining a suitable reward function for complex tasks can be challenging.

That's where human feedback comes into play. In RLHF, an AI agent learns from rewards and penalties provided by humans, rather than a predefined reward function. This approach combines the power of ML algorithms with the intuition, experience, and knowledge of human experts.

The process involves the following steps:

1. The AI agent interacts with the environment and takes action.

2. Human observers assess the agent's actions and provide feedback in the form of rewards or penalties.

3. The agent uses this feedback to update its learning and improve its decision-making over time.

Through these interactions, the AI agent learns to perform complex tasks by incorporating human guidance.

There are several benefits to using RLHF. Primarily, human feedback allows the AI agent to learn from the wealth of knowledge and experience that humans possess. This approach enables agents to learn complex behaviors that may be difficult to achieve with traditional algorithms. At the same time, the learning process can be tailored to specific needs or goals by adjusting the feedback provided by human experts.

The drawbacks of having humans in the loop are that humans can make mistakes or provide inconsistent feedback, which may affect the agent's learning. Furthermore, training an AI agent using human feedback can be a slow process, as it requires continuous input from human experts. In other words, it tends to be labor intensive and potentially costly as a result. For this reason, it's important to develop an upfront estimate of the human and financial resources required for such a project to ensure its viability.

It's important to note that SMEs can be incredibly valuable contributors to almost any ML exercise. For example, we often use SMEs to help us review the outputs of our models because it allows us to discover new contexts in the problem space that should become features in the training data.

In the example of missing verification codes, we discovered the root of the problem by first developing a deep knowledge of the specific failure point by interviewing call center staff (one type of SME) and listening to call recordings (becoming SMEs ourselves). Once we had narrowed down the possible issue, we dug into the inner workings of the core system with our colleagues from the IT department (another type of SME) to verify the glitch.

This approach is not the same as reinforcement learning, but it highlights the value of involving SMEs throughout the whole development process, even if it requires manual input.

Codifying labeling rules with programmatic labeling

The traditional method of labeling using human annotators is sometimes a bottleneck in the process that can prevent us from creating high-quality training sets in a way that is both efficient and cost-effective. This is typically an issue when we're dealing with large datasets. Time- and cost-efficient training has become increasingly important as ML models become more complex and datasets become larger.

Enter programmatic labeling. At its core, programmatic labeling is a process of automatically assigning labels to data points based on predefined rules or algorithms. The main advantage of programmatic labeling over manual labeling is that it can be done much faster and more accurately than manual labeling – once there is a robust labeling function in place. This makes it ideal for large datasets where manual labeling would take too long or be too costly.

The process of programmatic labeling begins with defining the labels that need to be assigned to each data point. This can be done manually by SMEs or through automated methods such as **natural language processing (NLP)** algorithms or rule-based systems. Once the labels have been defined, they can then be applied to the data points using either supervised or unsupervised learning algorithms.

The main benefits of programmatic labeling are as follows:

- **Scalability**: Programmatic labeling can handle large volumes of data more efficiently than manual labeling, enabling faster model training and iteration

- **Consistency**: Automated labeling methods ensure a consistent application of rules and criteria across the entire dataset, reducing variability and potential errors that may arise from human subjectivity

- **Cost-effectiveness**: By automating the labeling process, organizations can save on the time and resources required to train, manage, and compensate human labelers

- **Speed**: Programmatic labeling can process and annotate data much more quickly than manual labeling, accelerating the overall ML pipeline

- **Reduced human error**: Automation minimizes the risk of human error and inconsistencies that can be introduced during manual labeling

- **Reproducibility**: The automated labeling process is easily replicable, ensuring that results can be reproduced and verified across different datasets and projects

- **Adaptability**: Programmatic labeling algorithms can be fine-tuned and updated as needed to accommodate changing requirements, new data sources, or evolving project goals

- **24/7 availability**: Unlike human labelers, programmatic labeling can operate continuously without breaks or downtime, allowing for uninterrupted progress in ML projects

We will show you how to use specific programmatic labeling techniques in *Chapter 6, Techniques for Programmatic Labeling in Machine Learning*.

Programmatic labeling techniques are often all that's required to lift the quality of your data. However, in some cases, relationships between features are too complex for rules-based algorithms to do the job. This brings us to the third principle of data-centric ML.

Principle 3 – use ML to improve your data

Just as we can use a programmatic or algorithmic approach to label our data, we can also use ML to identify data points that may be wrong or ambiguous. By leveraging developments in explainability, error analysis, and semi-supervised approaches, we can create new labels and find data points to improve or discard.

Here are some practical steps to generate better input data with ML:

- **Toss out noisy examples**: Sometimes, more data is not always better. Noisy data can lead to inaccurate predictions. By removing noisy examples, we can improve the quality of our input data. For instance, if you're analyzing customer reviews and some reviews are filled with random characters or irrelevant information, those can be considered as "noisy" and removed.

- **Use techniques to focus on a subset of data to improve**: Not all data has the same value. We can focus on a subset of data to improve the quality of our input data. For example, if you're analyzing sales data, you might focus on the subset of data from your most profitable region to get the most return on your efforts, all else being equal.

- **Expand available label data by leveraging ML generalization from expert input**: ML can be used to expand the available label data by using expert input to achieve similar precision and greater coverage. An expert in bird species, for example, could provide input on a limited set of images, and ML could use this to accurately label a larger set of images.

- **Use semi-supervised approaches**: Semi-supervised approaches, including weak learning and active learning, can be used to identify data points that require SME review. For example, you might use active learning to identify customer emails that need to be reviewed by a human for sentiment analysis.

- **Use explainability**: Explainability is essential in identifying patterns in data and ensuring that models make sense. Complex models require a model-specific or model-agnostic approach to explainability, including local and global methods and SHAP values. For example, using SHAP values can help you understand why your model predicted a certain outcome in a loan approval process, ensuring the decision-making is transparent and explainable.

- **Use error analysis**: Error analysis can help identify patterns in data where models are making mistakes, helping to improve the quality of our input data. For instance, if your model is incorrectly identifying cats as something else in image recognition, error analysis can help you figure out where and why it's making these mistakes.

The techniques required to perform these steps will be outlined throughout the subsequent chapters of this book.

By applying these steps in production, we can identify performance drifts in labeling functions or models. Additionally, we can identify data points that require human review, leading to better quality input data and improved prediction accuracy.

The use of ML to improve input data quality is a fundamental shift in the traditional approach to ML. It requires a mindset shift from using ML models to make the best prediction to using ML to identify the data points that are not helping model performance. After all, the goal of data-centric ML is to increase signal and reduce noise in our input data.

Embracing a data-centric approach also provides us with a unique opportunity to collect and refine data in a manner that is inherently aligned with ethical, responsible, and well-governed ML practices. This shift in focus allows us to design our data strategies not just around performance enhancement but also around principles of fairness, transparency, and accountability.

As we proceed, we will explore how this approach can help us to embed ethics at the very core of our data collection and refinement processes. This way, we can ensure that improved data quality goes hand in hand with maintaining the integrity and trustworthiness of our ML applications.

Principle 4 – follow ethical, responsible, and well-governed ML practices

Ethical and responsible ML practices become increasingly important as data-centricity allows us to tackle more high-stakes challenges. This requires you to consider factors such as transparency, fairness, and accountability when designing algorithms so that they do not discriminate against certain groups or individuals. Additionally, those responsible for implementing these systems must be aware of how they work and understand their limitations so that they can make informed decisions about their use.

Unfortunately, ethical and responsible ML practices are generally not as developed as they should be. In 2021, the IBM Institute for Business Value and Oxford Economics conducted a study[1] where 75% of executives ranked AI ethics as important; however, fewer than 20% of executives strongly agreed that their organizations' practices aligned with their declared principles and values.

As practitioners of data-centric ML, we need to consider that the term *data quality* is much broader than the objective accuracy of individual data points. High-quality data should also allow us to identify and monitor potential ethical issues throughout the ML development process and beyond.

AI ethics and responsibility is not just a tick-box exercise, but a potential source of differentiation. Organizations that pay attention to AI ethics are more likely to be trusted by their customers, while organizations that overlook it are likely to suffer customer backlash and reputational damage[2].

The story of the UK's 2020 *school grading fiasco* highlights what can happen when you ignore ethical considerations while using ML in high-stakes environments. During the COVID-19 pandemic, students across the UK were unable to sit their exams because of lockdowns. Instead, an algorithm was used to grade students' exam results, resulting in a significant number of students receiving lower grades than they deserved. This caused uproar among students, teachers, and the academic community as it was seen as unfair and unjust.

The algorithm used by Ofqual, the UK regulator responsible for regulating qualifications, exams, and assessments, was designed to standardize grades across different schools to make them comparable. It considered factors such as prior attainment and school performance. However, it did not consider individual student performance or teacher assessments, which resulted in many students receiving lower grades than they should have.

Instead, the model favored students from private institutions and wealthy areas, significantly impacting high-performing individuals from public, state-funded schools. Consequently, numerous students lost their university admissions due to the lowered exam scores. This caused a great deal of distress among the students who had worked hard for their exams only to be let down by an algorithm that did not accurately reflect their abilities. In the end, the grades awarded by the algorithm were canceled, and replaced by a fairer but more manual grading approach.

To avoid similar incidents such as the UK grading disaster from occurring in the future, AI systems must be designed with ethical considerations in mind from the outset. Overall, this incident highlights some of the ethical issues associated with AI systems and demonstrates why it is important for us to consider these issues when designing and implementing them. It also serves as a reminder that we must ensure these systems are transparent, fair, and accountable if we want them to be effective tools for decision-making in our society.

We will be discussing specific ways to deal with ambiguity in labeling in *Chapter 4, Data Labeling Is a Collaborative Process* and show you a range of techniques for identifying and removing bias in *Chapter 8, Techniques for Identifying and Removing Bias*.

Summary

In this chapter, we outlined the four principles of data-centric ML. By following these principles, you will be able to create ML models that are based on high-quality data that has been enhanced, cross-checked, and verified by humans, labeling functions, and ML techniques.

This allows us to get more signals out of our data, which, in turn, increases our ability to build powerful models on small or large datasets. Lastly, we can capture ethical considerations throughout the development life cycle, which ultimately ensures we're using our powers for good.

In the next chapter, we'll explore specific ways you can structure, optimize, and govern the process of using human annotators for your ML projects.

References

1. https://www.ibm.com/thought-leadership/institute-business-value/en-us/report/ai-ethics-in-action, accessed on 1 June 2023

2. https://www.capgemini.com/insights/expert-perspectives/decoding-trust-and-ethics-in-ai-for-business-outcomes/, accessed on 1 June 2023

4

Data Labeling Is a Collaborative Process

As the field of **artificial intelligence** (**AI**) continues to evolve, publicly available tools such as ChatGPT, **Large Language Model Meta AI** (**LLaMA**), Bard, Midjourney, and others have set a new benchmark for what's possible to achieve with structured and unstructured data.

These models obviously rely on advanced algorithms and massive amounts of data, but many people are unaware that human labeling remains a critical component in their ongoing refinement and advancement. As an example, ChatGPT's model infrastructure relies on individuals reviewing and annotating data samples that are then fed back into the model to improve its understanding of natural language and context.

In this chapter, we explore how to get the most out of data collection and annotation tasks involving human labelers. We will cover these general topics:

- Why we need human annotators
- Understanding common challenges arising from human labeling tasks
- Designing a framework for achieving high-quality labels
- Best-practice approaches for motivating human annotators, avoiding bias, and dealing with ambiguity in labeling

Firstly, let's understand why human input is a cornerstone of data-centric **machine learning** (**ML**).

Understanding the benefits of diverse human labeling

Incorporating a diverse range of individuals and perspectives in the human labeling process offers several advantages. Humans bring a level of precision and accuracy to data annotation that is difficult for machines to match. While automated systems may struggle with ambiguity or complexity, human annotators can leverage their understanding and reasoning capabilities to make informed decisions.

Data can change over time, and new scenarios can arise that were not present in the original training data. Human annotators can adapt to these changes, providing updated annotations that reflect the new realities. This ensures that ML models remain relevant and effective as the data evolves.

Some key strengths of human labelers over programmatic labeling include the following:

- **Domain expertise**: Labelers with subject-matter expertise can provide valuable insights and annotations that help the model better comprehend specific topics and domains.

- **Active learning**: This approach involves prioritizing data samples that the model finds ambiguous or challenging, enabling labelers to focus on areas where their input can have the greatest impact.

- **Diversity of perspectives**: A diverse group of labelers can help mitigate potential biases in the training data, leading to a fairer and more inclusive AI model. By involving labelers from diverse backgrounds and with varied experiences, a model can be exposed to a broader range of linguistic nuances, cultural contexts, and perspectives, improving its overall performance.

- **Enhanced contextual understanding**: By drawing on the experiences and knowledge of labelers from different backgrounds, the model can develop a deeper understanding of language nuances, idioms, and cultural references. Exposure to a wide variety of perspectives and inputs can make the model more resilient and versatile, enabling it to handle a broader range of tasks and scenarios effectively.

- **Quality control (QC)**: Regular audits and evaluations of labeler output can help ensure consistent annotation quality and adherence to guidelines, which is essential for effective model training.

- **Adherence to ethics**: There may be data or scenarios that shouldn't end up as model input based on ethical considerations. In these cases, human labelers play a crucial role in helping models meet ethical standards. As an example, OpenAI, the company behind ChatGPT, uses human annotators to review, label, and filter out toxic "not safe for work" data.

Although human annotation is a key part of data-centric model development, humans also add to any ML project new behaviors, biases, and risks that must be managed. We will now discuss these typical challenges before presenting our framework for managing them.

Understanding common challenges arising from human labelers

Before we dive into the best practices of labeling accuracy and consistency, we will define common challenges we must tackle through our labeling framework. Labeling inaccuracy and ambiguity are generally triggered by one or more of the following seven causes:

- **Poor instructions**: Labeling inconsistencies will arise from unclear or insufficient instructions for the data annotation task. If annotators are not given clear guidelines, they may make assumptions or guesses that lead to inconsistent or inaccurate annotations.

- **Human bias**: Bias can introduce ambiguity when the data is skewed toward a particular result or outcome, leading to inaccurate interpretations. A common solution is to assign multiple annotators to label the same data, choosing the most frequently occurring label as the correct one. However, this aggregation or voting method can sometimes exacerbate bias rather than rectify it. For instance, if the majority of annotators have a particular bias, their consensus may reflect this bias rather than the true data.

- **Human error**: Annotators are humans, and humans make mistakes. Even the most well-trained, engaged, and focused annotators are one typo or mouse click away from applying an incorrect label leading to random noise in a dataset. These mistakes do occur but are unlikely to happen in a systematic way. Nevertheless, we need to have a way of identifying and correcting these mistakes so that they don't introduce unnecessary random noise.

- **Objective versus subjective tasks**: Every task lies on a spectrum from purely objective (with a single correct answer) to highly subjective (with many potentially correct *interpretations*). The more subjective a task is, the more ambiguity it tends to introduce into data annotation as different annotators may have different interpretations. As you will learn in this chapter, even tasks that seem relatively straightforward can contain hidden layers of subjectivity at the boundary.

- **Difficulty**: Tasks that are inherently complex or hard to comprehend can lead to ambiguity in data annotation. If a task is too difficult, annotators may struggle to understand or complete it correctly, leading to inconsistent or inaccurate annotations.

- **Ambiguity**: Some tasks or datasets are naturally ambiguous, meaning there's room for multiple valid interpretations. This ambiguity can lead to inconsistencies in data annotation, as different annotators may interpret the same data in different ways.

- **Static versus variable labeling**: In static labeling, each data point is assigned a single, unchanging label. In contrast, variable labeling allows labels to change based on context or additional information. Variable labeling can introduce ambiguity as the same data point may be labeled differently in different contexts. This form of labeling inconsistency may also arise as annotators become more familiar with a task, which causes them to alter their perception of the definition of labels.

We will now introduce our framework for achieving accurate and consistent labels. It is specifically designed to identify or prevent issues stemming from these seven common labeling challenges.

Designing a framework for high-quality labels

Annotations and reviews done by humans can be labor-intensive and susceptible to human errors and inconsistency. As such, the goal is to build datasets that are both accurate and consistent, requiring labels to meet accuracy standards as well as ensuring results from different annotators are within the same range.

These goals may seem obvious at first, but in reality, it can be very tricky to get human labelers to conform to the same opinion. On top of that, we also need to verify that a consensus opinion is not biased somehow.

Our framework for achieving high-quality human annotations consists of six dimensions. We will briefly summarize these dimensions before delving into a detailed explanation of how to achieve them:

- **Clear instructions**: To ensure high-quality labels, the instructions for the annotation task must be explicit and unambiguous. The annotators should have a clear understanding of what is expected of them, including details about the task, the criteria for labeling, and examples of correctly labeled data.

- **Aligned motivations**: The annotators' motivations should align with the goal of obtaining high-quality labels. This could involve rewarding accuracy, providing feedback, and creating an environment that encourages meticulous work. When annotators feel that their work is valuable and recognized, they are more likely to produce high-quality labels.

- **Subject-matter experts (SMEs)**: Utilizing annotators who are experts in the subject matter can significantly improve the quality of labels. These individuals possess deep knowledge and understanding of the context behind the data, enabling them to recognize subtleties and nuances that others may miss.

- **Iterative collaboration**: High-quality labels can be achieved through a process of iterative collaboration. Annotators should be encouraged to communicate and collaborate, revisiting and refining their labels based on collective feedback and discussion.

- **Diversity of thinking**: A diverse group of annotators brings different perspectives and interpretations to the task, which can lead to more comprehensive and robust labels. Diversity of thinking can help uncover blind spots and reduce bias in the labeling process.

- **Dealing with ambiguity**: As ambiguity is inherent in any data annotation task, training annotators on how to handle ambiguous cases is essential for achieving high-quality labels. This could include strategies such as seeking additional information, consulting with peers or supervisors, or following predefined rules for ambiguous cases.

Let's delve into these six dimensions in detail to understand how to establish an exceptional labeling process. It all begins by ensuring our instructions to annotators are crystal clear.

Designing clear instructions

It may seem blatantly obvious that a labeling task should come with clear instructions, but as you will learn throughout this chapter, that is not necessarily an easy task. Assignments should be clear not just to the people creating them but also, more importantly, to the annotators who will execute them.

This challenge has three components to it: firstly, instructions should contain specific details for tasks to be carried out reliably, regardless of who is performing them. Secondly, the instruction design should include ways of picking up instruction issues early and throughout the assignment. Thirdly, we must make sure our annotators are adequately qualified and motivated for the task.

McInnis et al. (2016)[1] studied the issues that can arise if we don't manage one or more of these components. The researchers looked at the impact of unclear instructions and misaligned motivations between requestors and annotators using **Amazon Mechanical Turk (AMT)**.

They found that seasoned annotators use various tools and techniques to assess the quality of an assignment and the reliability of the requesters who posted it, before taking on new projects. They do this to pick clearly defined assignments that can be performed with no hiccups while avoiding tasks that will pay them little money or impact their reputation.

Basically, good annotators will circumvent unclear assignments to avoid iterative tasks or unfair rejection of completed work (rejection of an annotator's work results in non-payment for the work performed). The authors found that annotators commonly look for the following risk factors in task instructions when deciding whether to take on an assignment:

- Flaws in the task or interface design
- Unclear evaluation criteria
- Unresponsive, arbitrary resolution of rejections
- Lack of information on requesters
- Inexperienced and unfamiliar requesters
- Tasks with poor return
- Prioritizing efficiency over quality

As data professionals, it is our job to provide task instructions that mitigate these seven factors, whether we are using crowdsourced annotators or in-house SMEs. However, sometimes you won't know whether your instructions are clear until you use them in a live setting.

Therefore, the question is: how can we best align the understanding of a task between requestors and annotators while also ensuring that we have the right people on the job?

Liu et al. (2016)[2] developed a best-practice method for this purpose called **Gated Instruction**. This technique is used for training annotators, aligning the understanding of tasks between requestors and annotators, and identifying underperforming workers.

Gated Instruction is based on the idea that humans learn better when they are given feedback on their performance. This is done by providing an interactive teaching environment where users can receive feedback on their annotations and adjust their approach accordingly. The goal is to create a system that can provide accurate annotations with minimal effort from the user.

The Gated Instruction Crowdsourcing Protocol is a simple and generalizable three-phase process designed to ensure quality data annotation. It includes an interactive tutorial, screening questions, and batches of questions with continued screening. The authors describe the protocol as follows:

Phase I – Interactive tutorial

This phase involves a comprehensive tutorial that explains the task at hand:

1. Workers are given clear definitions of each relation and tagging criteria.
2. Workers practice by annotating sentences that illustrate each relation.
3. Immediate feedback is provided after each practice sentence to guide the workers.

Phase II – Screening questions

This phase is designed to evaluate the worker's understanding of the task:

1. Workers are asked to annotate a representative set of five gold standard questions.
2. Feedback is given on each question to help the workers understand their errors.
3. Workers who fail a majority of these questions are excluded from the remaining process.

Phase III – Batches of questions (with continued screening)

This phase focuses on maintaining high-quality work during the task execution:

1. Gold standard questions are included in the task without providing feedback.
2. Sets of five gold standard questions are included in each batch of 20 questions, with their frequency decreasing exponentially.
3. Workers who score less than 80% accuracy on the last 10 gold standard questions are eliminated.

General principles

These principles ensure the integrity and efficiency of the process:

1. Only workers with an AMT reputation above a certain threshold are accepted.
2. A link to the definitions of relations is provided throughout the task for quick reference.
3. Workers must correct any mistakes highlighted in the feedback before proceeding.
4. After each batch, feedback is provided on earnings so far and performance on gold standard questions.
5. Workers are reminded of a bonus upon completion of all 10 batches, encouraging consistent high-quality work.

Gated Instruction provides more accurate results than conventional instruction and screening methods since users are able to receive feedback on their annotations and adjust their approach accordingly. Using this approach, the authors improved precision from 0.50 to 0.77 and recall from 0.70 to 0.78 on the same dataset. This was in comparison to results from workers who were instructed using more traditional methods of instruction.

These two studies have used crowdsourced annotators to conduct their research, but the same frameworks are still highly relevant when your labelers are not crowdsourced workers, but SMEs with whom you have a closer working relationship.

For instance, if you're selecting annotators from a pool of in-house colleagues, you might base your selection on someone's particular expertise, years of experience, and interest in contributing to the project. As you will learn in the next section, annotators' motivation to perform a task can have a big influence on how data is collected and, ultimately, how your models perform on that data.

Let's build on these principles as we discuss how to motivate annotators and use SMEs for more complex labeling assignments.

Aligning motivations and using SMEs

While technology has made leaps and bounds in automating data collection, human data collectors still hold an essential place in various fields. They bring a level of understanding, empathy, and judgment that machines cannot replicate. For instance, human data collectors and annotators can interpret nuances in language, context, and emotions. Similarly, they can engage with respondents, build rapport, and encourage more open and honest responses.

A significant challenge with human data collectors is maintaining their motivation and engagement in alignment with secondary uses of this data. Let's discuss four factors that commonly contribute to this issue.

#1 – Lack of purpose

If data collectors do not understand the significance of their work, they may feel disconnected and unmotivated. A data collector might ask themselves, "Why am I collecting this information if no one is going to use it?"

Solution: Communicate the big picture and show how every data point becomes a valuable building block of ML solutions.

We start any data collection and labeling exercise with the assumption that workers are intelligent and able to understand the importance of collecting rich, unbiased information. They understand that any compromise in data quality can have negative implications for its future use. Our role then becomes to articulate the importance of good data collection and explain its intended use. This articulation will significantly mitigate issues related to data quality.

A couple of years ago, Manmohan and Jonas ran a number of ML projects that required front-line staff members to collect a bunch of information through customer interviews. These interviews had been running for years so we already had lots of data collected, but the information was captured as text in a conversational format and therefore difficult to interpret statistically.

We wanted to make this data easier to use for ML purposes, so we gathered all front-line workers for a presentation and workshop on the importance of data quality. After showing these data collectors how we used the data they collected to build specific ML solutions, they were surprised to learn how impactful their jobs could be.

As one team member said, "If only I had known that the information I collect from one person could be used to help thousands of other customers, I would have put way more effort into the details."

In the workshop, we made data collectors take ownership of the situation by creating a series of templated questions that would improve the accuracy and signal of collected data. Our front-line colleagues took great pride in improving the way they could collect data for the greater good of the company and its customers. They had a transformed sense of ownership for data quality and the results that came from "their data."

We were happy too as the uplift in data quality ended up boosting our model accuracy by up to 40% in some instances.

#2 – Tedious nature of data tasks

Data collection and labeling can be repetitive and monotonous, leading to boredom and disengagement. This creates an underlying incentive to complete tasks with minimal effort.

Solution: We use the following four strategies to reduce the boring parts of data collection and labeling: refining our questions, simplifying the process, eliminating unnecessary data, and automating wherever possible.

It is sometimes mindboggling how much impact you can have by teaching data collectors to ask better questions. More specific and targeted questions will typically yield more meaningful and engaging responses. This improves data quality while making the data collection process more interesting and less repetitive.

Simplifying the process could mean streamlining workflows, using more user-friendly software, or providing clear instructions and training to those involved in data collection. By making the process more straightforward, we can reduce the cognitive load on individuals and make the task less tiresome.

If you're like most data professionals, the notion of discarding or limiting data collection may induce a degree of anxiety. However, it's vital to appreciate that data collection shouldn't mean hoarding every bit of data you come across, especially if it gets in the way of maintaining a great user experience. In fact, collecting unnecessary or irrelevant data can add to the monotony of the task and create clutter that hinders data analysis. Therefore, it's crucial to identify and focus only on the data that's truly relevant to your research question or business objective.

Our last simplification strategy, automation, is often a great solution to the tedium of data collection, but we prefer to only introduce automation once we have exhausted the other three strategies. There is no point in automating something that should be done differently or not at all.

For example, scraping tools can automatically extract large volumes of data from websites or documents, and rules-based logic or ML algorithms can label and organize data. You will learn some of these automation techniques in the coding chapters of this book.

A little while back, we worked with a large business to build real-time predictions of a caller's propensity to buy investment products. The business wanted their call center staff to have this prediction served up to them during the call so that they could match callers with the right experts for their investment needs.

These screening calls would typically take 15 to 20 minutes to complete as call center staff painstakingly moved through pages and pages of questions relating to the caller's personal details, demographics, investment experience, and risk appetite.

Although lots of data were being collected, it was not the right kind of information to build our prediction on. We identified a small handful of supplementary data points required to reliably determine someone's product needs. However, adding these to the data collection process as additional questions was unfeasible because the screening calls were already long and tedious. To get the information we needed, we had to create a win-win scenario for callers, call center staff, and data scientists alike.

As we workshopped this challenge with the call center team, we discovered that about 20% of the existing questions could be enhanced to make the conversation more concise, while another 15% of questions could be removed entirely because they captured irrelevant or duplicate information. Lastly, 10% of the data collected could be prefilled or derived based on callers' answers to other questions.

After implementing these changes, the average call-handling time decreased by about 30% even after we added five new questions to the process. This reduction in call duration enabled the processing of more calls per day, leading to an increase in conversions into paying customers – a win-win for everyone involved.

#3 – Lack of incentives

Without appropriate rewards or recognition, data collectors may lack the motivation to perform at their best. For instance, if a person is incentivized by the *quantity* of data points collected instead of the *quality* of individual responses, it may result in more superficial completion of tasks.

Solution: We have seen the best results when we focus on motivating data collectors across four different areas:

1. **Show what good looks like**: Firstly, it's essential to set clear expectations for what constitutes high-quality work. Most people take pride in their work and strive to perform well when they understand the value and impact of their efforts. By defining the purpose and significance of the data collection or labeling task and demonstrating examples of excellent work, you provide

a tangible target for your team to aim for. This clarity helps ensure the collected data is robust, relevant, and rich in valuable insights while minimizing noise.

2. **Create mutual benefits**: Secondly, fostering a sense of shared success between data collectors and data users is crucial. When data collectors understand how their efforts contribute to the bigger picture – perhaps driving key business decisions or fueling innovative projects – they are more likely to feel invested in their work. This sense of ownership and contribution can significantly enhance the quality of data collection.

3. **Provide non-monetary rewards**: The third area of focus is non-monetary rewards. Recognition and appreciation are powerful motivators. Gamification can be a highly effective tool in this regard, transforming the process of data collection into an engaging competition. Implementing features such as publicly displayed leaderboards, badges, or points can foster a sense of achievement and encourage healthy competition among team members. The longstanding concept of "Employee of the Month" not only rewards exceptional performance but also sets a standard for others to aspire to.

4. **Provide monetary rewards**: Finally, monetary rewards can be a potent incentive. Tying compensation to the quality of work can drive individuals to meet or exceed set standards. However, this approach requires a clear framework outlining performance expectations and objective measures of performance. At the same time, you also need to have the ability to influence someone's compensation, which can be difficult if your annotators and data collectors are in-house staff. While not always feasible, monetary incentives can be a powerful motivator when available. If they are not an option, doubling down on the other three areas of motivation can still yield impressive results.

In conclusion, motivating data collectors and annotators to do a great job is multifaceted. It requires a mix of clear communication, mutual benefits, recognition, and appropriate compensation.

#4 – Work environment

A stressful or unsupportive work environment can also affect motivation levels. If a data collector or annotator has many competing priorities, they may adopt a "close enough is good enough" attitude toward data collection.

Solution: Creating a suitable working environment for annotators and data collectors to focus on the main task at hand basically requires you to remove any unnecessary tasks or friction getting in the way of someone performing their job.

We regularly refuse to work on data science projects unless everyone needed for the project is motivated to participate and makes themselves available for the duration of the project. This often requires us to negotiate for a dedicated time in the calendar with SMEs, data collectors, and their managers.

It is also important to provide a feedback mechanism for data collectors. That way, they can contribute to the process and call out stuff that doesn't work or could be done better, such as questions that could be framed differently, changing the order of questions or tasks, confusing labeling requirements, and so on.

Give annotators and data collectors the necessary tools and resources to do their job swiftly and effectively. Examples include digitizing data collection, automating data population wherever possible, conditionally formatting drop-down options, and so on.

Continuous and timely feedback between project participants is also vital. Surprising workers with negative feedback late in the process can be demotivating and counterproductive. Instead, collaborate and provide feedback throughout the process to ensure alignment from the outset. Assume that there may be hiccups along the way, and be ready to coach your team through them proactively. Conversely, if you assume that everyone has understood every instruction as intended, then you're setting yourself up for failure.

Some people don't consider this kind of work part of the data science domain. In our opinion, this is where a lot of data professionals fall short. It's important to recognize that human behavior and bias can be a seriously limiting factor for data science projects, and you want to reduce this impact as much as possible. If you don't lean into these non-technical hurdles, you will end up with poorer data as a result.

This brings us to the next dimension in our framework: iterative collaboration.

Collaborating iteratively

Iterative collaboration should be a central strategy for human labeling tasks. Basically, it entails creating an ongoing process of feedback and fine-tuning the data labeling process. Here are three guiding principles for implementing a collaborative approach to data labeling, validated by data labeling platform SUPA's best-practice approach[3].

Start small and iron out any issues early

Initiating the data labeling process with smaller datasets is a practical approach. Instead of starting with thousands of observations, begin with a calibration batch of, say, 50 observations. This manageable dataset allows you to review the labels, spot potential problems, enhance instructions, and provide feedback to the labelers.

Repeat this process until you are confident that you have gone through a representative sample of the full dataset, that you have identified most edge cases, and that your labelers can perform the task consistently.

Labeling inconsistencies across the same observations indicates a need for rule revision. If annotators are interpreting guidelines differently, then it's worth understanding whether the cause is unclear labeling rules, discrepancies in the understanding and experience of individual labelers, or something entirely different. In other words, your labeling instructions are never set-and-forget. They should be under constant supervision and evolve to meet the needs of the project.

Visualize what good looks like

Annotation rules can occasionally be vague, leading to subjective interpretations and inconsistencies. For instance, in an image labeling exercise, a rule such as "Only label an item when most of it is visible" might be interpreted differently by different labelers. Therefore, it's important to visually show what good looks like.

Visual illustrations are invaluable in the data labeling process, as they offer labelers explicit guidelines on how labeled objects should appear. We recommend showing examples of how to perform a labeling task correctly, but also examples of the opposite. By offering visual illustrations of good and bad labeling conventions, labelers gain a deeper understanding of the task, thereby enhancing their productivity and precision.

Be very specific about edge cases

Edge cases are situations that deviate from the ordinary and can result in inconsistent labels due to subjective opinions. For instance, should a toy car be labeled as a car? Is a tandem bicycle one or two bicycles, or is it in its own category?

To manage edge cases effectively, you need to have a mechanism in place for annotators to flag these items for further consideration. If you don't have a feedback mechanism in place, it's likely that annotators will make up their own judgment on the spot to complete the task.

Pradhan et al. (2022)[4] propose a so-called *FIND-RESOLVE-LABEL* workflow for crowdsourced annotation, aimed at addressing these three iteration steps. The FIND-RESOLVE-LABEL workflow is a guided labeling process designed to reveal ambiguities for a specific labeling task and associated instructions.

For instance, a labeling task might be to identify images with a woman in it. On the surface, this task appears rather simple, but in reality, annotators can quickly face ambiguity. For example, when is someone a woman, and when is someone a girl? Does it even matter? What about a statue of a woman or the *Mona Lisa* painting – does that count? What if the woman is only partially visible?

In reality, it can be very difficult to provide high-quality instructions for a given labeling task because there can be so many dimensions to consider.

The FIND-RESOLVE-LABEL workflow aims to discover and remove these ambiguities at the beginning of the labeling exercise. It consists of three key components that work together to streamline the data labeling process:

1. **Find**: In this initial stage, annotators are provided with labeling instructions and asked to identify examples that are ambiguous based on these instructions. For each identified example, labelers are also asked to provide a concept tag that provides an explanation for why a certain label was chosen. This allows for the collection of the rationale and conceptual thinking behind labeling decisions, which can then be fed back into improved labeling instructions.

2. **Resolve**: Once the data points are identified, the next step involves resolving any ambiguities or conflicts in the data. This may require domain expertise to make informed decisions on how to address inconsistencies or missing information.

3. **Label**: Finally, after resolving any issues, the data points are labeled appropriately, ensuring high-quality annotations that can be used for training ML models.

Pradhan and his team discovered that focusing on the most unclear data points and clarifying them during labeling greatly improved data quality. They noticed that in some ambiguous scenarios, many annotators agreed on answers that were different from what the requester regarded as correct. This means that even when smart answer aggregation methods are used, there's a risk of getting incorrect labels for these tasks.

Interestingly, the study also found that workers could correctly label observations closely related to the main concept. This suggests we might not need to explain every possible ambiguity during the task because a well-chosen set of examples could help the team correctly label other unclear examples.

With these findings in mind, let's explore how to deal with ambiguity among annotators in a data-centric fashion.

Dealing with ambiguity and reflecting diversity

With the help of human annotators, we can produce datasets that are incredibly rich in information, but this sometimes requires us to tackle ambiguity in innovative ways. At the same time, ambiguity can be hard to spot. What one person finds obvious, another person may find entirely confusing.

Companies and researchers use internal staff, volunteers, or crowdsourcing platforms such as AMT to gain access to human annotators at affordable rates. These labelers come from diverse backgrounds and carry different biases, all of which can impact the quality of labeling – especially when there is some element of judgment involved.

This challenge only grows as AI and ML are used to classify and generate new content from datasets that can be interpreted differently depending on context and who is doing the interpretation. This is demonstrated in a research paper, *The Risk of Racial Bias in Hate Speech Detection*, by Sap et al. (2019)[5], which investigates how annotators' insensitivity to differences in dialect can lead to racial bias in automated hate speech detection.

The paper states that even datasets designed specifically for detecting hate speech contain an inherent bias toward specific groups or minority languages. This is because the underlying parameters are created based on the preferences of annotators who may be unaware of subtle nuances between different ethnicities or languages.

For example, some ethnicities or social groups may use colloquial language that seems rude or offensive to people from other demographics. For instance, the researchers discovered that labelers tended to mark phrases in African American English as being more toxic than those using General American English.

These inherent biases are difficult to avoid because annotators are rarely complete SMEs but people following general instructions. At the same time, labelers are unlikely to be a representative sample of the general public. For example, the majority of AMT participants have historically been comprised of younger individuals who are unmarried and without children. The vast majority of Turkers hail from only two countries, the US and India, while less than 2% come from the Global South[6].

This problem doesn't just pertain to the subject of hate speech. As such, certain dialects, lifestyles, cultural backgrounds, and worldviews may be overrepresented, while others remain underrepresented when determining any kind of label that requires subjective interpretation. The resulting poor data collection can lead to an increased gap between the labels used and the real-world scenarios they are meant to represent.

The imbalance that exists in data labeling is evident in some of the world's most widely used public training datasets. Research has found that two of the most common databases, *ImageNet* and *Open Images*, are biased toward the US and Europe, evident by the fact that models created from these datasets have poorer performance on images sourced from the Global South[7].

Images of grooms, for instance, receive a lower accuracy rating when they come from Ethiopia or Pakistan compared to similar images from the US. This particular discrepancy is due to the way objects such as "wedding" and "spices" are interpreted depending on their cultural context, with publicly available recognition systems struggling to correctly classify them when sourced from countries outside of America or Europe.

Understanding approaches for dealing with ambiguity in labeling

Labels are often absolute, but opinions are not. At the same time, humans will tend to disagree on what more abstract labels should be. As data-centric practitioners, we should anticipate ambiguity and disagreement among annotators and have a plan in place for managing it.

It's worth noting that we actually *want* ambiguity to show up so that we can deal with it in the right way. Ambiguous scenarios can arise because labeling instructions are unclear, but they can also uncover new labels that must be included in our dataset. Therefore, we should try to design our labeling teams to maximize the likelihood that we will tease out disagreements if they are there.

A great way to do this is to involve a more diverse pool of annotators who are better attuned to differences in elements such as language and opinion, but in doing this, we also want to elevate the opinions of minority groups. Stanford University researchers Gordon et al. (2022)[8] propose an approach for this purpose, called *Jury Learning*.

In previous research by Gordon et al. (2021)[9], it was found that when accounting for labels from non-majority groups in a comment toxicity labeling task, the classifier's performance decreased from 0.95 ROC AUC to 0.73 ROC AUC. This means that the classifier is not as effective when applied to comments from people who are not part of the majority group. In other words, it is impossible to make everyone agree, so we should consider who we listen to – not necessarily the simple majority.

Jury Learning stands in contrast to more straightforward aggregation or majority voting methods. Rather than basing labels on a majority rule or a probability, Jury Learning actively uses varying opinions to pick out underlying biases and suppress minority opinions. It is proposed as a way to integrate dissenting voices into ML systems in order to prevent them from becoming over-reliant on a single opinion or view. Here is how it works.

Jury Learning is a **supervised ML (SML)** approach that resolves disagreements explicitly through the metaphor of a jury. This approach allows practitioners to specify whose voices their classifiers reflect, and in what proportion. The goal of Jury Learning is to define which people or groups determine a system's prediction and in what proportion, allowing developers to analyze – and potentially mitigate – any potential biases that may be present in the model.

In order to use Jury Learning effectively, practitioners must first identify the jurors they wish to include in their model. This can be done by selecting individuals who represent different perspectives on the problem at hand.

For example, let's say the task is to label whether a movie is good. Practitioners could select jurors (labelers) from different demographics such as age, gender, ethnicity, location, political orientation, or socio-economic status. Once the jurors have been selected, they must then provide input into the model's predictions by answering questions about whether a movie is good and why.

Once all of the jurors have provided their input into the model's predictions, practitioners can then use this data to create an aggregate prediction for each individual case. By considering multiple perspectives on each case, practitioners can ensure that their models are more accurate and less biased than if they had relied solely on one perspective when making predictions.

Finally, practitioners can also use Jury Learning to analyze any potential biases present in their models by comparing the aggregate predictions made by different juries, composed of individuals from different backgrounds or perspectives. This comparison effectively provides us with a prediction range rather than binary labels. The analysis can help identify any areas where bias may be present and allow practitioners to adjust their models accordingly in order to reduce any potential bias and improve overall accuracy. This process is illustrated in the following diagram:

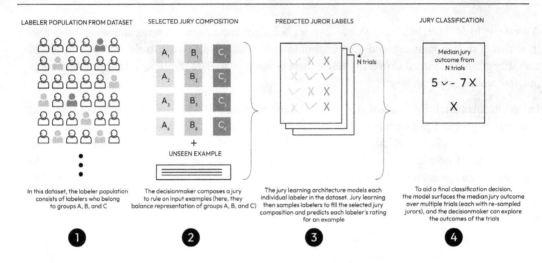

LABELER POPULATION FROM DATASET   SELECTED JURY COMPOSITION   PREDICTED JUROR LABELS   JURY CLASSIFICATION

In this dataset, the labeler population consists of labelers who belong to groups A, B, and C

The decisionmaker composes a jury to rule on input examples (here, they balance representation of groups A, B, and C)

The jury learning architecture models each individual labeler in the dataset. Jury learning then samples labelers to fill the selected jury composition and predicts each labeler's rating for an example

To aid a final classification decision, the model surfaces the median jury outcome over multiple trials (each with re-sampled jurors), and the decisionmaker can explore the outcomes of the trials

①   ②   ③   ④

Figure 4.1 – An overview of the Jury Learning process taken from Gordon et al. (2022)

By now, you may have noticed that Jury Learning is not simple to set up. As with many data-centric approaches, Jury Learning requires extensive planning and coordination to ensure that all participants have a clear understanding of the expectations and guidelines of the process. This can be a time-consuming and resource-intensive process.

Jury Learning also requires a labeler pool of a certain size to ensure the necessary diversity of opinions and views. This can be a challenge, particularly for projects with limited resources.

Another challenge that comes with implementing Jury Learning is the need for meta-information on labelers. To ensure that the results are both accurate and reliable, Jury Learning requires labelers with diverse skill sets and backgrounds. Gathering this information and developing a pool of labelers that meets the required standards can be a difficult task that again requires upfront planning.

Despite these challenges, Jury Learning provides an innovative and effective way for practitioners to incorporate dissenting voices into ML models while also helping them identify and mitigate any potential biases present in their models. By considering multiple perspectives when making predictions and analyzing potential biases present in their models, practitioners can ensure that their ML models are both accurate and unbiased when making decisions about important tasks that are prone to subjectivity.

To round off this chapter, let's explore how we can measure ambiguity or disagreement among annotators statistically.

Measuring labeling consistency

So far, we have discussed a range of tools and techniques for creating consistent and high-quality annotations. While these elements create the foundation for good datasets, we also want to be able to measure whether our annotators are performing consistently.

To gauge annotator consistency, we recommend using two measures of labeling consistency called intra- and interobserver variability, respectively. These are standard terms in clinical research and refer to the degree of agreement among different measurements or evaluations made by the same observer (intra-) or by different observers (inter-). To simplify the explanation, consider "observer" to be interchangeable with "labeler," "annotator, "rater," "data collector," and any other similar term we have used throughout this chapter.

While both intra- and interobserver variability relate to measurement consistency, they address different aspects. Intra-observer variability refers to the consistency of a single observer over time, while inter-observer variability refers to the consistency between different observers. Factors such as training, experience, and standardization of protocols can significantly influence both.

Tracking observer variability is crucial as it directly impacts the quality and reliability of your input dataset and, therefore, your model. If the same object is interpreted differently by various observers (inter-observer variability) or even by the same observer at different times (intra-observer variability), it could lead to inconsistencies in labeling, thereby affecting the overall quality of ML outputs.

Several factors contribute to observer variability, including lack of standardization in measurement techniques, observer fatigue, and subjective interpretations. As an example, someone's judgment might be influenced by their level of experience, personal bias, or even their state of mind at the time of observation.

It's important to note that labeling discrepancies between two or more observers do not necessarily mean that one observer is correct and others are incorrect. Labeling disagreements may simply mean that the labeled object or situation is ambiguous or transitory and therefore difficult to give a hard label.

For instance, two doctors might disagree on the relative progression of an illness based on medical imaging or symptom descriptions. This could be because the diagnosis is uncertain rather than one doctor being more correct than the other.

A common technique for measuring variability is the **intraclass correlation coefficient (ICC)**. ICC is a statistical tool used to assess the consistency or conformity of annotations made by one or more observers measuring the same entity. Unlike the commonly used Pearson correlation coefficient, which measures linear relationships between variables, ICC assesses the reliability of ratings within the same group of data. It's particularly useful when we want to know how strongly units in the same group resemble each other.

A high ICC value close to 1 indicates a high similarity between values from the same group. Conversely, a low ICC suggests less agreement among the ratings.

There are different forms of ICC, each applicable in specific circumstances. For instance, some forms are more suitable when we have a single measurement from each subject, while others are better suited for an average of several measurements. The choice of form depends on the nature of your study and the type of data you have.

The following screenshot shows six common definitions outlined by Shrout and Fleiss (1979)[10]:

ICC Type	Definition	Formula
ICC1	This type refers to a situation where every subject is evaluated by a *different set of randomly chosen* observers. Reliability is determined from one isolated measurement. The formula measures absolute agreement.	$\dfrac{MS_R - MS_W}{MS_R + (k+1)MS_W}$
ICC2	In this model, each subject is evaluated by every observer, and these observers are viewed as a sample from a larger group of similar observers. The reliability stems from a single measurement.	$\dfrac{MS_R - MS_E}{MS_R + (k+1)MS_E} + \dfrac{k}{n}(MS_C - MS_E)$
ICC3	ICC3 refers to a scenario where a specific group of observers rate each subject. There's no extension to a broader group of observers. Both ICC2 and ICC3 adjust for average differences among reviewers (notice that they're calculated using the same formula), but they're influenced by interactions between variables. The distinction between ICC2 and ICC3 lies in whether we view the reviewers as a constant or variable factor.	$\dfrac{MS_R - MS_E}{MS_R + (k+1)MS_E}$
ICC1k ICC2k ICC3k	Similar to the above, but in this instance, reliability is calculated by taking an average of the 'k' observers' measurements.	$ICC1k = \dfrac{MS_R - MS_W}{MS_R}$ $ICC2k = \dfrac{MS_R - MS_E}{MS_R + \dfrac{MS_C - MS_E}{n}}$ $ICC3k = \dfrac{MS_R - MS_E}{MS_R}$

Variable Definitions

- MS_R (Mean Square for Rows): measures the variability between the different groups or categories. In ICC it usually represents the variability between subjects.
- MS_W (Mean Square Within): measures the variability within each group or category. It accounts for the error or residual variance that can't be explained by the model.
- MS_E (Mean Square Error): estimates the variance within groups. MS_E is specifically about the discrepancy between a model's predictions and the actual outcomes, while MS_W is about the variance within groups that can't be accounted for by group-level differences.
- MS_C (Mean Square Columns): measures the variability between the different raters or measurements.
- n: The total number of subjects being studied or evaluated.
- k: The number of ratings per subject (or the number of raters).

Figure 4.2 – Six common ICCs

To build our intuition around the use of ICC scores, let's work through a practical example using the **Pingouin** Python package. Pingouin is an open source package with a large number of useful statistical features. It primarily utilizes pandas and NumPy, so make sure you have these installed as well.

For our example scenario, suppose we have four wine-tasting judges rating the quality of eight different wines by giving them a score of 0 to 9. We would like to know whether these judges are rating the wines consistently. The judges' ratings are displayed in the following screenshot:

Wine #	Judge A	Judge B	Judge C	Judge D
1	1	2	0	1
2	1	3	3	2
3	3	8	1	4
4	6	4	3	3
5	6	5	5	6
6	7	5	6	2
7	8	7	7	9
8	9	9	9	8

Figure 4.3 – Wine-tasting scores from four different judges

- Generally speaking, three types of variability can occur:

- Variability due to differences in the objects being assessed (suppose two different samples of the same wine had slightly different tastes)

- Variability caused by the assessment of observers, for example, the difference between judges B and C's rating of wine *#3*

- Variability in the use of labels; for example, everyone finds wine *#1* the worst, but three different ratings have been used to rate it

The ICC calculation will take all of these into account as it is based on **analysis of variance (ANOVA)** analysis. Before we get to calculating ICC, we must first determine which type of measure we're after. We can use the following screenshot as a guide to selecting the correct form of ICC for our situation:

Are all subjects evaluated by the same group of observers?	Is our observer sample randomly chosen from a larger group, or is it a specific selection?	Is focus on the reliability of one observer's score or the average score from multiple observers?	What level of agreement are we looking for in the observers' scores?	Model
No	Yes	Single	Absolute	ICC1
Yes	Yes	Single	Absolute	ICC2
Yes	No	Single	Consistency	ICC3
No	Yes	Multiple (k)	Absolute	ICC1k
Yes	Yes	Multiple (k)	Absolute	ICC2k
Yes	No	Multiple (k)	Consistency	ICC3k

Figure 4.4 – ICC model selection guide

In our scenario, the following applies:

- Each judge has rated each wine once, so we can determine that all subjects have been evaluated by the same group of observers

- In this case, we will assume that the four judges have been chosen randomly from a larger pool of potential judges

- We are interested in the reliability of individual observers rather than the average reliability

Therefore, we're looking to use the ICC2 calculation to determine our reliability score.

We then use the following script to run the ICC function over our wine scores. The Pingouin ICC operator, `intraclass_corr`, will calculate and present all six common ICC measures, but we are only interested in ICC2:

```
import pingouin as pg
data = pg.read_dataset('icc')
icc = pg.intraclass_corr(data=data, targets='Wine', raters='Judge',
                         ratings='Scores').round(3)
icc.set_index("Type")
```

This produces the following output. Our ICC score is 0.728, which means our judges are in moderate agreement on ratings:

Type	Description	ICC	F	df1	df2	pval	CI95%
ICC1	Single raters absolute	0.728	11.680	7	24	0.0	[0.43, 0.93]
ICC2	Single random raters	0.728	11.787	7	21	0.0	[0.43, 0.93]
ICC3	Single fixed raters	0.729	11.787	7	21	0.0	[0.43, 0.93]
ICC1k	Average raters absolute	0.914	11.680	7	24	0.0	[0.75, 0.98]
ICC2k	Average random raters	0.914	11.787	7	21	0.0	[0.75, 0.98]
ICC3k	Average fixed raters	0.915	11.787	7	21	0.0	[0.75, 0.98]

Figure 4.5 – Output table

It's important to understand that there are no strict thresholds for what constitutes an "acceptable" ICC score. Though there are no rigid benchmarks, some general guidelines can help interpret ICC scores. These ranges are not absolute and should be interpreted in the context of your specific study or analysis:

- **ICC less than 0.5**: This range is generally considered to indicate poor reliability. For instance, if you have an ICC of 0.3 for a set of ratings, it would suggest a low level of agreement among raters.

- **ICC between 0.5 and 0.75**: Scores in this range are typically considered to show moderate reliability.

- **ICC between 0.75 and 0.9**: These scores suggest good reliability. If you achieve an ICC of 0.8, for example, it indicates a high degree of agreement among your raters.

- **ICC greater than 0.90**: This range represents excellent reliability. An ICC of 0.95, for example, would suggest almost perfect agreement among raters.

When interpreting ICC scores, it's also important to consider several factors that can influence their reliability. These include the following:

- **Sample size**: As with many statistical measures, the ICC is sensitive to sample size. Larger sample sizes tend to provide more reliable estimates of the ICC.

- **Data outcome range**: ICC scores can also be impacted by the potential range of outcomes being assessed and the difficulty of determining annotations. For instance, if our wine judges could only label wines as "good" or "bad" (1, 0) rather than a range (0–9), then that would likely alter the final ICC score.

- **Subject variability**: The ICC is also influenced by the variability among subjects. High subject variability can lead to lower ICC values, even when raters are consistent in their ratings.

In practice, interpreting ICC scores requires understanding the context of your study, the nature of your data, and the specific form of ICC used. Always consider these factors when interpreting and communicating your results.

Remember – ICC scores are just one piece of the puzzle. They should be used in conjunction with other statistical measures and insights to provide a comprehensive understanding of your data. Let's summarize what we've covered so far in the chapter.

Summary

Throughout this chapter, we've examined the critical role that humans play in ensuring data quality, particularly in the initial stages of data labeling. We've recognized that while human labelers are indispensable, they also present certain challenges, including biases and inconsistencies.

To address these issues, we've explored various strategies to train labelers effectively for high-quality dataset development. The key takeaway here is that well-trained labelers, armed with clear instructions, can significantly increase the overall quality of your data.

Improving task instructions emerged as a recurring theme, underscoring their importance in facilitating the labeling process. Iterative collaboration was also highlighted as an essential practice, promoting continuous improvement through feedback and refinement.

By the end of this chapter, you should have gained a comprehensive understanding of why human involvement is crucial in data-centric model building, the challenges posed by human labelers, and practical ways to overcome them. More importantly, you'll have learned how to use specific frameworks to achieve quality labeling, setting a solid foundation for successful ML projects.

In the next chapter, we will build on these skills and delve deeper into the technical aspects of data cleaning and augmentation before we explore programmatic labeling techniques in *Chapter 6, Techniques for Programmatic Labeling in Machine Learning*. It's time to get deep into code!

References

1. *McInnis B., Cosley D., Nam C., Leshed G., Taking a HIT: Designing around Rejection, Mistrust, Risk, and Workers' Experiences in Amazon Mechanical Turk, Information Science & Law School, Cornell University.* `https://dl.acm.org/doi/epdf/10.1145/2858036.2858539`

2. *Liu A., Soderland S., Bragg J., Lin C. H., Ling X., Weld D. S., Effective Crowd Annotation for Relation Extraction, Turing Center, Department of Computer Science and Engineering, University of Washington.* `https://aclanthology.org/N16-1104.pdf`

3. `https://www.supa.so/post/iteration-a-key-data-labeling-process-often-overlooked`, viewed July 30, 2023.

4. *Pradhan V. K., Schaekerman M., Lease M., 2022, In Search of Ambiguity: A Three-Stage Workflow Design to Clarify Annotation Guidelines for Crowd Workers, Front. Artif. Intell., 18 May 2022, Sec. Machine Learning and Artificial Intelligence Volume 5 - 2022.* `https://doi.org/10.3389/frai.2022.828187`

5. *Sap, M., Card, D., Gabriel, S., Choi, Y., Smith, N. A., The Risk of Racial Bias in Hate Speech Detection, Paul G. Allen School of Computer Science & Engineering, University of Washington, Seattle, USA, Machine Learning Department, Carnegie Mellon University, Pittsburgh, USA, Allen Institute for Artificial Intelligence, Seattle, USA*

6. `https://venturebeat.com/business/the-ai-industry-is-built-on-geographic-and-social-inequality-research-shows/`, viewed April 22, 2023.

7. `https://venturebeat.com/ai/mit-researchers-find-systematic-shortcomings-in-imagenet-data-set/`, viewed April 22, 2023.

8. *Gordon M. L., Lam M. S., Park J. S., Patel K., Hancock J., Hashimoto T., Bernstein M. S., 2022. Jury Learning: Integrating Dissenting Voices into Machine Learning Models. In CHI Conference on Human Factors in Computing Systems (CHI '22), April 29-May 5, 2022, New Orleans, LA, USA. ACM, New York, NY, USA*

9. *Gordon M. L., Zhou K., Patel K., Hashimoto T., Bernstein M.S., 2021. The Disagreement Deconvolution: Bringing Machine Learning Performance Metrics In Line With Reality. In CHI Conference on Human Factors in Computing Systems (CHI '21), May 8-13, 2021, Yokohama, Japan. ACM, New York, NY, USA, 14 pages.* `https://doi.org/10.1145/3411764.3445423`

10. *Shrout, P. E. & Fleiss, J. L. (1979), Intraclass correlations: uses in assessing rater reliability, Psychological Bulletin, 86(2), 420.* `https://psycnet.apa.org/doi/10.1037/0033-2909.86.2.420`, viewed July 30, 2023.

Part 3:
Technical Approaches
to Better Data

In this part, we explore technical approaches to enhance data quality and management in machine learning. We cover topics ranging from data cleaning, programmatic labeling, and synthetic data usage, to addressing bias and handling rare events. Each chapter gives you essential skills and knowledge to work efficiently with data in machine learning, highlighting how important good quality data is in building robust ML systems.

This part has the following chapters:

- *Chapter 5, Techniques for Data Cleaning*
- *Chapter 6, Techniques for Programmatic Labeling in Machine Learning*
- *Chapter 7, Using Synthetic Data in Data-Centric Machine Learning*
- *Chapter 8, Techniques for Identifying and Removing Bias*
- *Chapter 9, Dealing with Edge Cases and Rare Events in Machine Learning*

Techniques for Data Cleaning

In this chapter, we will cover six key dimensions of data quality and their corresponding techniques to improve data quality, commonly known as techniques for cleaning data in machine learning. Simply put, data cleaning is the process of implementing techniques to improve data quality by fixing errors in data or removing erroneous data. As covered in *Chapters 1* and *2*, reducing errors in data is a highly efficient and effective way to improve model quality over using model-centric techniques such as adding more data and/or implementing complex algorithms.

At a high level, data cleaning techniques involve fixing or removing incorrect, incomplete, invalid, biased, inconsistent, stale, or corrupted data. As data is captured at multiple sources, due to different annotators following their judgment or due to poor system designs, combining these sources can often result in data being mislabeled, inconsistent, duplicated, or incomplete. As discovered in earlier chapters, incorrect data can make algorithms and outcomes unreliable. So, to achieve reliability in machine learning systems, it is important to help data scientists and data engineers question the data and systematically improve data quality using data cleaning techniques.

The topics that will be covered in this chapter are as follows:

- The six key dimensions of data quality
- Measuring data quality
- Data cleaning techniques required to improve data quality across the six dimensions

The six key dimensions of data quality

There are six key dimensions we can use to check the overall health of data. Ensuring good health across the data can ensure we can build reliable systems and make better decisions. For example, if 20% of survey data is duplicated, and the majority of the duplicates are filled by male candidates, we can imagine that the actions taken by decision-makers will favor the male candidates if data duplication is undetected. Hence, it's important to understand the overall health of the data to make reliable and unbiased decisions. To measure data quality or look at the overall health of the data, we can break down data quality into the following dimensions:

- **Consistency**: This refers to whether the same data is maintained across the rows for a given column or feature. An example of this could be whether the gender label for males is consistent or not. The label can take values of "1," "Male," "M", or "male," but if the data has multiple values to represent males, then an algorithm will treat each label individually, which can cause randomness and errors. Hence, the goal of ensuring consistency is to ensure labels are defined consistently across the whole dataset.

- **Uniqueness**: This refers to whether each record can be uniquely identified. If duplicate values enter the system, the model will become biased due to those records. For example, if one region has a high loan approval rate and due to system failure, records from this region are duplicated, the algorithm will be biased toward approving more loans from that region.

- **Completeness**: This refers to whether data is complete across the rows for a given column or feature, and whether data is missing due to system errors or not captured, especially when that information will be used in the machine learning system.

- **Validity**: This refers to whether the data labels conform to the rules. For example, if a label is present in the data but cannot be verified by any external source, this can be referred to as invalid data. In a housing loan dataset, the location of the property may have not conformed to the rules, where `semi_urban` might be invalid if one or a couple of annotators believed some suburbs are neither urban nor rural, and they violated the rules of data and entered `semi_urban`. This can introduce noise in the data, so it's important to ensure data conforms to the business rules.

- **Accuracy**: This refers to whether data was entered correctly in the first place and can be verified from an internal or external source. An example of this in a healthcare setting could be that if the date and time of admission are entered in a different time zone, then the analysis and insights would be inaccurate. If the time of admission is a predictor of quality of care, then misaligned time zones might lead to wrong conclusions.

- **Freshness**: This refers to whether the data available is recent and up to date to meet data requirements. Some applications require data to be real-time – that is, updated every second – whereas other applications require data to be available once a month or every few months. What was true yesterday may change due to changes in factors such as regulation, weather conditions, trends, competition, business changes, and more.

Next, we will install various Python packages, and then dive into fixing and measuring data quality. In each section, we dive deeper into different data cleaning techniques and how to improve the quality of data.

Installing the required packages

For this chapter, we will need the following Python packages or libraries:

- `pandas` version 1.5.3
- `numpy` version 1.22.4
- `scikit-learn` version 1.2.1
- `jupyter` version 1.0.0
- `alibi` version 0.9.0
- `alibi-detect` version 0.10.4
- `seaborn` version 0.12.2
- `matplotlib` version 3.6.3
- `missingno` version 0.5.1
- `feature-engine` version 1.5.2

Next, we will provide a brief introduction to the dataset and start exploring the data.

Introducing the dataset

First, let's introduce our problem statement. For loan providers, it is important to ensure that people who get a loan can make payment and don't default. However, it is equally important that people are not denied a loan due to a model trained on poor-quality data. This is where the data-centric approach helps make the world a better place – it provides a framework for data scientists and data engineers to question the quality of data.

For this chapter, we will use the loan prediction dataset from Analytics Vidhya. You can download the dataset from `https://datahack.analyticsvidhya.com/contest/practice-problem-loan-prediction-iii/#ProblemStatement`. There are two files – one for training and one for testing. The test file doesn't contain any labels. For this chapter, we will utilize the training file, which has been downloaded and saved as `train_loan_prediction.csv`.

First, we will look at the dataset and check the first five rows. To do this, we must import the following necessary packages:

```
import pandas as pd
import numpy as np
import missingno as msno
import matplotlib.pyplot as plt
```

Next, we will read the data:

```
df = pd.read_csv('train_loan_prediction.csv')
df.head().T
```

We will read the data using pandas' `read_csv` method. Then, we will visualize the first five rows of the dataset using `.head()`. We can apply the `.T` method at the end of the `.head()` method if we have a large feature set. This will represent columns as rows and rows as columns, where column names will not exceed 5 as we want to visualize the first five rows.

We get the following output:

	0	1	2	3	4
Loan_ID	LP001002	LP001003	LP001005	LP001006	LP001008
Gender	Male	Male	Male	Male	Male
Married	No	Yes	Yes	Yes	No
Dependents	0	1	0	0	0
Education	Graduate	Graduate	Graduate	Not Graduate	Graduate
Self_Employed	No	No	Yes	No	No
ApplicantIncome	5849	4583	3000	2583	6000
CoapplicantIncome	0.0	1508.0	0.0	2358.0	0.0
LoanAmount	NaN	128.0	66.0	120.0	141.0
Loan_Amount_Term	360.0	360.0	360.0	360.0	360.0
Credit_History	1.0	1.0	1.0	1.0	1.0
Property_Area	Urban	Rural	Urban	Urban	Urban
Loan_Status	Y	N	Y	Y	Y

Figure 5.1 – The first five rows of our df dataset

As we can see, there are some inconsistencies across column names. All the columns follow the camel case convention, except `Loan_ID`, and long column names are separated by _ except for `LoanAmount`, `CoapplicantIncome`, and `ApplicantIncome`. This indicates that the column names have inconsistent naming conventions. Within the data, we can also see that some columns have data in camel case but `Loan_ID` has all its values in uppercase. Within the `Education` column, the `Not Graduate` value is separated by a space. In machine learning, it's important to ensure data is consistent; otherwise, models may produce inconsistent results. For instance, what happens if the `Gender` column has two distinct values for male customers – `Male` and `male`? If we don't treat this, then our machine learning model will consider `male` data points separate from `Male`, and the model will not get an accurate signal.

Next, we will extract the list of column names from the data, make them all lowercase, and ensure words will be separated by a unique character, _. We will also go through the data values of categorical columns and make them all lowercase before replacing all the special characters with _ and making our data consistent.

Ensuring the data is consistent

To ensure the data is consistent, we must check the names of the columns in the DataFrame:

```
column_names = [cols for cols in df]
print(column_names)
['Loan_ID', 'Gender', 'Married', 'Dependents', 'Education', 'Self_
Employed', 'ApplicantIncome', 'CoapplicantIncome', 'LoanAmount',
'Loan_Amount_Term', 'Credit_History', 'Property_Area', 'Loan_Status']
```

Next, we must get all the column names that don't contain underscores:

```
num_underscore_present_columns = [cols for cols in column_names if '_'
not in cols]
num_underscore_present_columns
['Gender',
 'Married',
 'Dependents',
 'Education',
 'ApplicantIncome',
 'CoapplicantIncome',
 'LoanAmount']
```

Since some columns have two uppercase letters in their names, we must add the underscore before the start of the second uppercase letter. Next, we create a Boolean mapping against the index of each letter of the column, where the location of capital letters will be mapped as True so that we can locate the index of the second capital letter and prefix it with an underscore:

```
cols_mappings = {}
for cols in num_underscore_present_columns:
    uppercase_in_cols = [val.isupper() for val in cols]
    num_uppercase_letters = sum(uppercase_in_cols)

    cols_mappings[cols] = {
        "is_uppercase_letter": uppercase_in_cols,
        "num_uppercase_letters": num_uppercase_letters,
        "needs_underscore": (num_uppercase_letters > 1)
    }
```

Then, we iterate over the mappings and print the column names that will require an underscore, and also print the location of the second capital letter:

```
for key in cols_mappings.keys():
    if cols_mappings[key]['needs_underscore']:
        print()
        print(f'{key} need the underscore at location ', cols_
mappings[key]['is_uppercase_letter'].index(True, 1))


ApplicantIncome need the underscore at location  9

CoapplicantIncome need the underscore at location  11

LoanAmount need the underscore at location  4
```

Using this information, we build some logic for the ApplicantIncome column:

```
'ApplicantIncome'[:9] + '_' + 'ApplicantIncome'[9:]
'Applicant_Income'
```

Next, we combine the previous steps and iterate over columns that require an underscore and build a mapping. Then, we print the names of the columns that require underscores. Finally, we create a mapping of old column names and new column names:

```
cols_mappings = {}
for cols in num_underscore_present_columns:
    uppercase_in_cols = [val.isupper() for val in cols]
    num_uppercase_letters = sum(uppercase_in_cols)

    if num_uppercase_letters > 1:
        underscore_index = uppercase_in_cols.index(True, 1)
        updated_column_name = cols[:underscore_index] + "_" +
cols[underscore_index:]
    else:
        updated_column_name = cols

    cols_mappings[cols] = {
        "is_uppercase_letter": uppercase_in_cols,
        "num_uppercase_letters": num_uppercase_letters,
        "needs_underscore": (num_uppercase_letters > 1),
        "updated_column_name": updated_column_name
    }
    if cols_mappings[cols]['needs_underscore']:
        print(f"{cols} will be renamed to {cols_mappings[cols]
['updated_column_name']}")
```

```
column_mappings = {key: cols_mappings[key]["updated_column_name"] for
key in cols_mappings.keys()}
column_mappings
```

```
ApplicantIncome will be renamed to Applicant_Income
CoapplicantIncome will be renamed to Coapplicant_Income
LoanAmount will be renamed to Loan_Amount
{'Gender': 'Gender',
 'Married': 'Married',
 'Dependents': 'Dependents',
 'Education': 'Education',
 'ApplicantIncome': 'Applicant_Income',
 'CoapplicantIncome': 'Coapplicant_Income',
 'LoanAmount': 'Loan_Amount'}
```

Finally, we apply the column mappings to update the column names and print the new column names:

```
df = df.rename(columns=column_mappings)
column_names = [cols for cols in df]
print(column_names)
['Loan_ID', 'Gender', 'Married', 'Dependents', 'Education', 'Self_
Employed', 'Applicant_Income', 'Coapplicant_Income', 'Loan_Amount',
'Loan_Amount_Term', 'Credit_History', 'Property_Area', 'Loan_Status']
```

Although the preceding code could have simply been updated by manually choosing the columns, by doing this programmatically, we can ensure that data is following programmatic rules. To improve consistency, we make all column names lowercase. First, we create some simple one-line logic to convert column names into lowercase:

```
print([cols.lower() for cols in df])
['loan_id', 'gender', 'married', 'dependents', 'education', 'self_
employed', 'applicant_income', 'coapplicant_income', 'loan_amount',
'loan_amount_term', 'credit_history', 'property_area', 'loan_status']
```

Then, we update the column names by passing the preceding logic:

```
df.columns = [cols.lower() for cols in df]
print(df.columns)
Index(['loan_id', 'gender', 'married', 'dependents', 'education',
       'self_employed', 'applicant_income', 'coapplicant_income',
       'loan_amount', 'loan_amount_term', 'credit_history', 'property_
area',
       'loan_status'],
      dtype='object')
```

With that, we have ensured that the column names are consistent. But more importantly, we must ensure that categorical values are consistent so that we can future-proof machine learning systems from inconsistent data. First, we extract the `id` and `target` columns and then identify the categorical columns. These columns contain non-numerical data:

```
id_col = 'loan_id'
target = 'loan_status'

cat_cols = [cols for cols in df if df[cols].dtype == 'object' and cols
not in [id_col, target]]
cat_cols
['gender',
 'married',
 'dependents',
 'education',
 'self_employed',
 'property_area']
```

We iterate over each column and check the unique values to ensure the values are distinct and not misspelled. We also check that the same value is not represented differently, such as it being in a different case or being abbreviated:

```
for cols in cat_cols:
    print(cols)
    print(df[cols].unique())
    print()
gender
['Male' 'Female' nan]

married
['No' 'Yes' nan]

dependents
['0' '1' '2' '3+' nan]

education
['Graduate' 'Not Graduate']

self_employed
['No' 'Yes' nan]

property_area
['Urban' 'Rural' 'Semiurban']
```

Looking at the data, it seems that values are distinct and not abbreviated or misspelled. But machine learning systems can be made future-proof. For instance, if we make all values lowercase, then in the future, if the same value comes with a different case, before entering the system, it will be made lowercase. We can also see that some strings, such as Not Graduate, take up space. Just like how we ensured consistency for column names, we must replace all white spaces with underscores. First, we create a new DataFrame named df_consistent; then, we make all categorical values lowercase and replace all spaces with underscores:

```
df_consistent = df.copy()
for col in cat_cols:

    df_consistent[col] = df_consistent[col].apply(lambda val: val.
lower() if isinstance(val, str) else val)

    df_consistent[col] = df_consistent[col].apply(lambda val: val.
replace(' ','_') if isinstance(val, str) else val)

for cols in cat_cols:
    print(cols)
    print(df_consistent[cols].unique())
    print()
gender
['male' 'female' nan]

married
['no' 'yes' nan]

dependents
['0' '1' '2' '3+' nan]

education
['graduate' 'not_graduate']

self_employed
['no' 'yes' nan]

property_area
['urban' 'rural' 'semiurban']
```

With that, we have ensured that data is consistent and all values are converted into lowercase.

As we can see, the dependents column contains numerical information. However, due to the presence of the 3+ value, the column values are encoded as strings. We must remove the special character and then encode this back to a numerical value since this column is ordinal:

```
df_consistent.dependents = df_consistent.dependents.apply(lambda val:
float(val.replace('+','')) if isinstance(val, str) else float(val))
```

Next, we look at the married and self_employed columns since these are binary and must be encoded to 1 and 0. The gender column has two values and can be binary encoded as well – for example, we can encode male as 1 and female as 0. The education column also has two values, and we can encode graduate as 1 and not_graduate as 0:

```
for cols in ['married', 'self_employed']:
    df_consistent[cols] = df_consistent[cols].map({"yes": 1, "no": 0})

df_consistent.education = df_consistent.education.map({
    'graduate': 1,
    'not_graduate': 0
})


df_consistent.gender = df_consistent.gender.map({
    'male': 1,
    'female': 0
})

for cols in cat_cols:
    print(cols)
    print(df_consistent[cols].unique())
    print()
gender
[ 1.  0. nan]

married
[ 0.  1. nan]

dependents
[ 0.  1.  2.  3. nan]

education
[1 0]

self_employed
[ 0.  1. nan]
```

```
property_area
['urban' 'rural' 'semiurban']
```

Now that the data is consistent and has been encoded correctly, we must create a function for preprocessing data so that we can consistently process any future variations to categorical labels. Then, we apply the function to the DataFrame and print the values to ensure the function was applied correctly:

```python
def make_data_consistent(df, cat_cols) -> pd.DataFrame:
    """Function to make data consistent and meaningful"""

    df = df.copy()

    for col in cat_cols:

        df[col] = df[col].apply(lambda val: val.lower() if
isinstance(val, str) else val)
        df[col] = df[col].apply(lambda val: val.replace(' ','_') if
isinstance(val, str) else val)


    df['dependents'] = df['dependents'].apply(lambda val: float(val.
replace('+','')) if isinstance(val, str) else float(val))

    for cols in ['married', 'self_employed']:
        df[cols] = df[cols].map({"yes": 1, "no": 0})

    df['education'] = df['education'].map({
        'graduate': 1,
        'not_graduate': 0
    })

    df['gender'] = df['gender'].map({
        'male': 1,
        'female': 0
    })


    return df
df_consistent = df.copy()
df_consistent = make_data_consistent(df=df_consistent, cat_cols=cat_
cols)

for cols in cat_cols:
    print(cols)
```

```
    print(df_consistent[cols].unique())
    print()
gender
[ 1.  0. nan]

married
[ 0.  1. nan]

dependents
[ 0.  1.  2.  3. nan]

education
[1 0]

self_employed
[ 0.  1. nan]

property_area
['urban' 'rural' 'semiurban']
```

We have now ensured that the data is consistent so that if categorical values have spaces instead of _ or are entered with a different case in the future, we can use the functionality we created here to clean the data and make it consistent before it enters our model.

Next, we will explore data uniqueness to ensure no duplicate records have been provided to create bias in the data.

Checking that the data is unique

Now that we have ensured the data is consistent, we must also ensure it's unique, before it enters the machine learning system.

In this section, we will investigate the data and check whether the values in the loan_id column are unique, as well as whether a combination of certain columns can ensure data is unique.

In pandas, we can utilize the .nunique() method to check the number of unique records for the column and compare it with the number of rows. First, we will check that loan_id is unique and that no duplicate applications have been entered:

```
df.loan_id.nunique(), df.shape[0]
(614, 614)
```

With this, we have ensured that loan IDs are unique. However, we can go one step further to ensure that incorrect data is not added to another loan application. We believe it's quite unlikely that a loan application will require more than one combination of income and loan amount. We must check that we can use a combination of column values to ensure uniqueness across those columns:

```
df[['applicant_income', 'coapplicant_income', 'loan_amount']].value_
counts().reset_index(name='count')
        applicant_income  coapplicant_income  loan_amount  count
0                   4333              2451.0        110.0      2
1                    150              1800.0        135.0      1
2                   4887                 0.0        133.0      1
3                   4758                 0.0        158.0      1
4                   4817               923.0        120.0      1
..                   ...                 ...          ...    ...
586                 3166              2985.0        132.0      1
587                 3167                 0.0         74.0      1
588                 3167              2283.0        154.0      1
589                 3167              4000.0        180.0      1
590                81000                 0.0        360.0      1

[591 rows x 4 columns]
```

As we can see, in the first row, there are two applications with the same income variables and loan amount. Let's filter the dataset to find these believed-to-be duplicate records by using values from the first row:

```
df[(df.applicant_income == 4333) & (df.coapplicant_income == 2451) &
(df.loan_amount == 110)]
        loan_id  gender married dependents education self_employed  \
328  LP002086  Female     Yes          0  Graduate            No
469  LP002505    Male     Yes          0  Graduate            No


        applicant_income  coapplicant_income  loan_amount  loan_amount_
term  \
328                 4333              2451.0        110.0         360.0
469                 4333              2451.0        110.0         360.0


        credit_history property_area loan_status
328                1.0         Urban           N
469                1.0         Urban           N
```

Looking at this subset, it is quite obvious that the data contains duplicates or two different applications were made – one by the husband and another one by the wife. This data is not providing any more information, other than that one application has been made by a male candidate and another has been made by a female candidate. We could drop one of the data points, but there is a big imbalance in the ratio of male to female applications. Also, if the second one was a genuine application, then we should keep the data point:

```
df.gender.value_counts(normalize=True)
Male      0.813644
Female    0.186356
Name: gender, dtype: float64
```

Based on this, we have understood what makes a data point unique – that is, a combination of `gender`, `applicant_income`, `coapplicant_income`, and `loan_amount`. It's our goal, as data scientists and data engineers, to ensure that once uniqueness rules have been defined, data coming into the machine learning system conforms to those uniqueness checks.

In the next section, we will discuss data completeness or issues with incomplete data, and how to handle incomplete data.

Ensuring that the data is complete and not missing

Now that we have achieved data consistency and uniqueness, it's time to identify and address other quality issues. One such issue is missing information in the data or incomplete data. Missing data is a common problem with real datasets. As a dataset's size increases, the chance of data points going missing in the data increases. Missing records can occur in several ways, some of which include:

- **Merging of source datasets**: For example, when we try to match records against date of birth or a postcode to enrich data, and either of these is missing in one dataset or is inaccurate, such occurrences will take NA values.

- **Random events**: This is quite common in surveys, where the person may not be aware of whether the information required is compulsory or they may not know the answer.

- **Failures of measurement**: For example, some traits, such as blood pressure, are known to have a very substantial component of random error when measured in the conventional way (that is, with a blood pressure cuff). If two people measure a subject's blood pressure at almost the same time, or if one person measures a subject's blood pressure twice in rapid succession, the measured values can easily differ by 10 mm/Hg (`https://dept.stat.lsa.umich.edu/~kshedden/introds/topics/measurement/`). If a person is aware of these errors, they may decide to omit this information; for some patients, this data will take NA values. In finance, an important measurement ratio to determine the credit worthiness of someone or a firm is the debt-to-income ratio. There are scenarios when income is not declared, and in those circumstances, dividing debt by 0 or missing data would result in missing information for the ratio.

- **Poor process design around collecting data**: For example, in health surveys, people are often asked about their BMI, and not everyone knows their BMI or understands the measurement. It would be simpler and easier if we were to ask for someone's height and weight as they are more likely to know this. Another problem arises when someone is asked about their weight measurement, where some people might be likely to omit or lie about this information. If BMI cannot be understood or measured at the time of collecting data, the data will take NA values.

When training datasets contain missing values, machine learning models can produce inaccurate predictions or fail to train properly due to the lack of complete information. In this section, we will discuss the following techniques for handling missing data:

- Deleting data

- Encoding missingness

- Imputation methods

One way to get rid of missing data is by deleting the missing records. This is also known as the **complete case analysis** (**CCA**) method. This is fine if less than 5% of rows contain missing values, but deleting more records could reduce the power of the model since the sample size will become smaller. There might also be a systematic bias in the data since this technique assumes that the data is missing completely and random, but it violates other assumptions such as when data is **missing at random** (**MAR**) or **missing not at random** (**MNAR**). Hence, blindly removing the data could make the model more biased. For instance, if a minority population has not declared their income in the past or has not held credit in the past, they may not have a credit score. If we remove this data blindly without understanding the reason for missingness, the algorithm could be more biased toward giving loans to majority groups that have credit information, and minority groups will be denied the opportunity, despite some members having solid income and creditworthiness.

Let's explore the dataset using the CCA technique, remove all rows where information is missing, and figure out what volume of data is lost:

```
remaining_rows = df_consistent.dropna(axis=0).shape[0]
total_records = df_consistent.shape[0]
perc_dropped = ((total_records - remaining_rows)/total_records)*100

print("By dropping all missing data, only {:,} records will be left
out of {:,}, a reduction by {:,.3f}%".format(remaining_rows, total_
records, perc_dropped))
By dropping all missing data, only 480 records will be left out of
614, a reduction by 21.824%
```

Since 21% is almost one-fourth of the dataset, this is not a feasible method. Hence, in this section, we will explore how to identify missing data, uncover patterns or reasons for data being missing, and discover techniques for handling missing data so that the dataset can be used for machine learning.

First, we will extract categorical features, binary features, and numerical features. To do this, we must separate the identifier and the target label:

```
id_col = 'loan_id'
target = 'loan_status'

feature_cols = [cols for cols in df_consistent if cols not in [id_col,
target]]
binary_cols = [cols for cols in feature_cols if df_consistent[cols].
nunique() == 2]
cat_cols = [cols for cols in feature_cols if (df_consistent[cols].
dtype == 'object' or df_consistent[cols].nunique() <= 15)]
num_cols = [cols for cols in feature_cols if cols not in cat_cols]
cat_cols
['gender',
 'married',
 'dependents',
 'education',
 'self_employed',
 'loan_amount_term',
 'credit_history',
 'property_area']
binary_cols
['gender', 'married', 'education', 'self_employed', 'credit_history']
num_cols
['applicant_income', 'coapplicant_income', 'loan_amount']
```

To check if data is missing in the dataset, pandas provides a convenience method called `.info()`. This method shows us how many rows are complete among the total records. The method also displays the data type of each column:

```
df_consistent.info()
<class 'pandas.core.frame.DataFrame'>
RangeIndex: 614 entries, 0 to 613
Data columns (total 13 columns):
 #   Column              Non-Null Count   Dtype
---  ------              --------------   -----
 0   loan_id             614 non-null     object
 1   gender              601 non-null     float64
 2   married             611 non-null     float64
 3   dependents          599 non-null     float64
 4   education           614 non-null     int64
 5   self_employed       582 non-null     float64
 6   applicant_income    614 non-null     int64
 7   coapplicant_income  614 non-null     float64
```

```
8    loan_amount           592 non-null    float64
9    loan_amount_term      600 non-null    float64
10   credit_history        564 non-null    float64
11   property_area         614 non-null    object
12   loan_status           614 non-null    object
dtypes: float64(8), int64(2), object(3)
memory usage: 62.5+ KB
```

The pandas library has another convenience method called .isnull() to check which row is missing data for a column and which row is complete. By combining .sum() with .isnull(), we can get the total number of missing records for each column:

```
df_consistent.isnull().sum()
loan_id                  0
gender                  13
married                  3
dependents              15
education                0
self_employed           32
applicant_income         0
coapplicant_income       0
loan_amount             22
loan_amount_term        14
credit_history          50
property_area            0
loan_status              0
dtype: int64
```

As we can see, the credit_history, self_employed, and loan_amount columns have the most missing data. Raw values can sometimes be difficult to comprehend and it's more useful to know the percentage of data that's missing from each column. In the next step, we will create a function that will take the DataFrame and print out the missing percentage of data against each column. Then, we sort the data in descending order of missingness:

```
def missing_data_percentage(df: pd.DataFrame):
    """Function to print percentage of missing values"""

    df = df.copy()

    missing_data = df.isnull().sum()
    total_records = df.shape[0]

    perc_missing = round((missing_data/total_records)*100, 3)
```

```
    missing_df = pd.DataFrame(data={'columm_name':perc_missing.index,
'perc_missing':perc_missing.values})


    return missing_df



missing_data_percentage(df_consistent[feature_cols]).sort_
values(by='perc_missing', ascending=False)
            columm_name  perc_missing
9          credit_history         8.143
4           self_employed         5.212
7            loan_amount         3.583
2             dependents         2.443
8       loan_amount_term         2.280
0                 gender         2.117
1                married         0.489
3              education         0.000
5        applicant_income         0.000
6      coapplicant_income         0.000
10          property_area         0.000
```

Now, we can extract the magnitude of missing data. However, before we dive into handling missing data, it is important to understand the patterns for missing data. By understanding these relationships, we will be able to take appropriate steps. This is because imputing missing data can alter the distribution of the data, which may further affect variable interaction.

We will utilize the missingno library and other visualizations to understand where data is missing, and in the absence of subject matter experts, we will make some assumptions on reasons for missing data.

To see where values are missing and where there are gaps in the data, we will utilize a matrix plot. A matrix plot can be quite useful when the dataset has depth or when the data contains time-related information. The presence of data is represented in gray, while absent data is displayed in white:

```
msno.matrix(df_consistent[feature_cols], figsize=(35, 15))
<AxesSubplot: >
```

Here's the output:

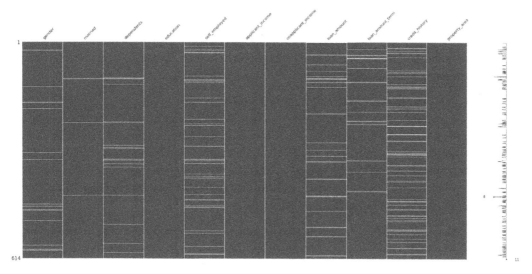

Figure 5.2 – Matrix plot

Looking closer, we can see that the `credit_history` column has a lot of missing points, and the occurrence of missingness is spread throughout the data and not at a given point in time.

As we touched upon earlier, understanding the reasons behind the missingness of data can help us choose the right technique to handle missing data. At a high level, we can call these mechanisms of missing data and classify them into three categories:

- **Missing completely at random (MCAR)**
- **Missing not at random (MNAR)**
- **Missing at random (MAR)**

Data is MCAR when the likelihood of missing data is the same for all the observations, and there is no relationship between the data being missing and any other features in the dataset. For example, a mail questionnaire might get lost in the post or a person may have forgotten to answer a question if they were in a hurry. In such cases, data being missing has nothing to do with the type of question, age group, or gender (relationship with other variables), and we can classify such features or data points as MCAR. Removing these data points or changing the value to 0 for such instances will not bias the prediction.

On the other hand, data is MAR when the likelihood of a data point being missing depends on other existing data points. For instance, if men don't disclose their weight 5% of the time on average, whereas women don't disclose their weight 15% of the time, we can assume that missingness in data is caused by the presence of gender bias. This will lead to a higher percentage of data being missing for women than for men. For this mechanism, we can impute data using statistical techniques or machine learning to predict the missing value by utilizing other features in the dataset.

The third mechanism, MNAR, can often be confused with MCAR but is slightly different. In this scenario, a clear assumption can be made as to why data is not missing at random. For instance, if we are trying to understand what factors lead to depression (outcome), depressed people could be more likely to not answer questions or less likely to be contacted. Since missingness is related to the outcome, these missing records can be flagged as "missing" and for numerical features, we can use a combination of machine learning to impute missing data from other features and flag data points, where data is missing, by creating another variable.

Now that we understand the different types of missing data, we will utilize the `heatmap` function from `missingno`, which will create a correlation heatmap. The visualization shows a nullity correlation between columns of the dataset. It shows how strongly the presence or absence of one feature affects the other.

Nullity correlation ranges from -1 to 1:

- -1 means if one column (attribute) is present, the other is almost certainly absent
- 0 means there is no dependence between the columns (attributes)
- 1 means that if one column (attribute) is present, the other is also certainly present

Unlike a standard correlation heatmap, the following visualization is about the relationship between missing data features since few of them have missing data. Those columns that are always full or always empty have no meaningful correlation and are removed from the visualization.

This heatmap helps identify data completeness correlations between attribute pairs, but it has limited explanatory ability for broader relationships:

```
msno.heatmap(df_consistent[feature_cols], labels=True)
```

This results in the following heatmap:

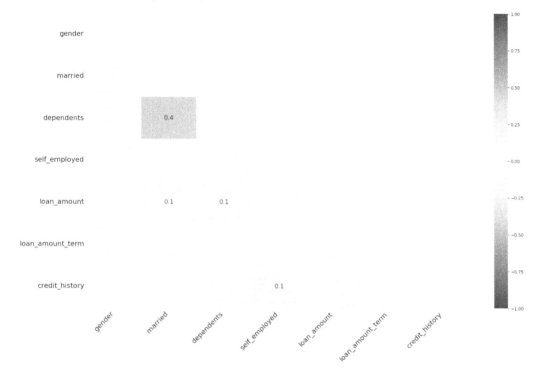

Figure 5.3 – Heatmap plot

From this plot, we can interpret relationships of missingness across a few variables. There is a correlation of 0.4 between `dependents` and `married`, which makes sense as the majority of the time, someone gets married first before having dependents.

Next, we will extract columns that contain missing data and use these for the next visualization. The `dendrogram` method uses hierarchical clustering and groups attributes together where missingness is associated with the missingness of another variable or completeness is associated with the completeness of another variable:

```
missing_cols = [cols for cols in feature_cols if df_consistent[cols].
isnull().sum() > 0]
msno.dendrogram(df_consistent[missing_cols])
```

The output is as follows:

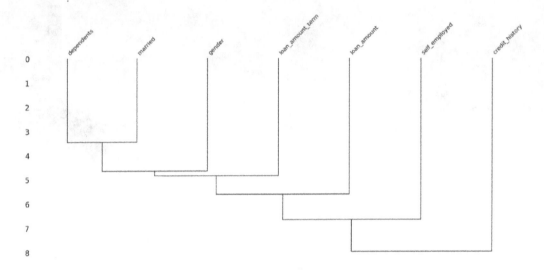

Figure 5.4 – Dendrogram plot

We interpret the dendrogram based on a top-down approach – that is, we focus on the height at which any two columns are connected with matters of nullity. The bigger the height, the smaller the relation, and vice versa. For example, the missingness or presence of data in credit_history has no relationship with the missingness or completeness of any other variable.

With that, we have understood patterns of missing data and if there are relationships between missing data columns. Next, we will explore the relationship between missing data and the outcome. Before we decide to remove missing data or impute it, we should also look at whether the missingness of a variable is associated with the outcome – that is, is there a chance that data may be MNAR?

First, we will visualize this relationship in missing categorical data:

```
cat_missing = [cols for cols in cat_cols if df_consistent[cols].
isnull().sum() > 0]

def cat_missing_association_with_outcome(data, missing_data_column,
outcome):
    """Function to plot missing association of categorical varibles
with outcome"""
```

```
    df = data.copy()
    df[f"{missing_data_column}_is_missing"] = df[missing_data_column].
isnull().astype(int)
    df.groupby([outcome]).agg({f"{missing_data_column}_is_missing":
'mean'}).plot.bar()

for cols in cat_missing:
    cat_missing_association_with_outcome(df_consistent, cols, target)
```

This will create some plots and show how categorical features are associated with the target variable:

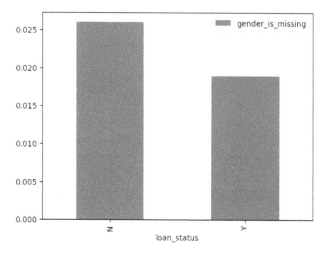

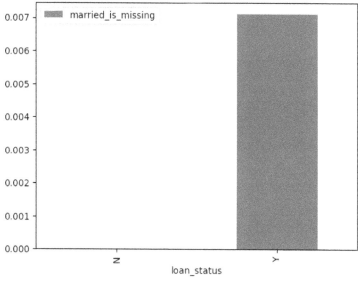

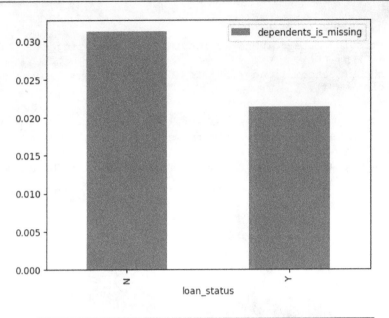

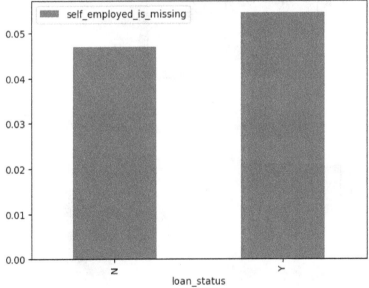

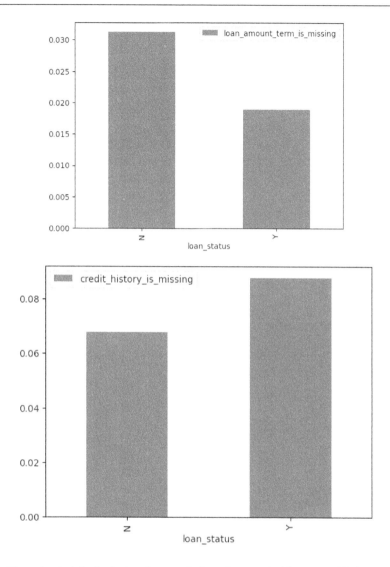

Figure 5.5 – The output plots displaying the association of categorical features with the target variable

At a high level, we can assume that for variables such as `married`, `dependents`, `loan_amount_term`, `gender`, and `credit_history`, the missingness of data is associated with the loan-approved status. Hence, we can say that the data for these variables is MNAR. For these three variables, we can encode missing data with the word "missing" as this signal will help predict the outcome. The missingness or completeness of `credit_history` is slightly associated with the `self_employed` status, as indicated in the heatmap plot, which shows that the data might be missing at random. Similarly, the missingness of the `married` status is associated with the missingness of `dependents` and `loan_amount`.

For all binary variables where data is missing, we can assume that data is not MCAR and rather assume that data is MNAR as there was some relationship of missingness of information with the outcome, or MAR, since missingness is associated with the presence or absence of other variables, as shown in the dendrogram.

One way to encode missing values is to encode these with the most frequent values or get rid of missing values or/and create an additional column that indicates missingness with 1 or 0. However, for MAR scenarios, this is not the best technique. As mentioned earlier, the goal of a data-centric approach is to improve data quality and reduce bias. Hence, instead of using frequency imputation methods or just deleting records, we should consider asking annotators to provide information where data is missing or make system fixes to recover from missing information. If that is not possible, we should consider using machine learning techniques or probabilistic techniques to determine possible values over simple imputation methods of mode, mean, and median. However, when missingness exceeds certain thresholds, even advanced techniques are not reliable and it's better to drop the feature. For the remaining variables, we will use a machine learning technique to determine the missing values since we cannot get annotators to help us provide complete information.

Now that we have identified how the missingness of categorical values is associated with the outcome, next, we will study the relationship between missing numerical data and the outcome:

```
num_missing = [cols for cols in num_cols if df_consistent[cols].
isnull().sum() > 0]


def num_missing_association_with_outcome(data, missing_data_column,
outcome):
    """Function to plot missing association of categorical varibles
with outcome"""

    df = data.copy()
    df[f"{missing_data_column}_is_missing"] = df[missing_data_column].
isnull().astype(int)
    df.groupby([outcome]).agg({f"{missing_data_column}_is_missing":
'mean'}).plot.bar()


for cols in num_missing:
    num_missing_association_with_outcome(df, cols, target)
```

This will display the following plot:

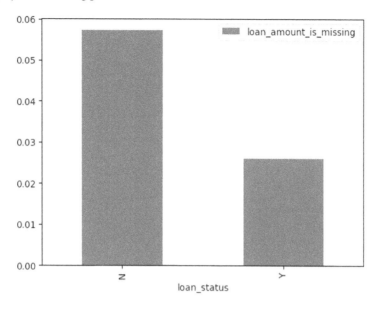

Figure 5.6 – Loan amount missing association with target

For `loan_amount`, it can be assumed that data is MNAR as well as MAR since the missingness or completion of data in the `married` and `dependents` variables is slightly associated with the missingness and completeness of `loan_amount`, as observed in the heatmap. Hence, we choose to impute missing values using machine learning and create an additional column to indicate missingness, which will provide a better signal to our model.

Next, we will dive into various methods of imputing data and compare these, as well as talking about the shortcomings of each approach. We will also discuss the implications of machine learning on imputing missing data in data-centric machine learning.

Following a model-centric approach, the standard rule of thumb for imputing numerical variables is that when 5% of the data is missing, impute it with the mean, median, or mode. This approach assumes that data is missing completely at random. If this assumption is not true, these simple imputation methods may obscure distributions and relationships within the data.

First, we will explore the distribution of `loan_amount` without imputation and when imputed with the median. The distribution changes when we impute 6% of the values with the median:

```
df_consistent.loan_amount.plot.kde(color='orange', label='loan_
amount', legend=True)
df_consistent.loan_amount.fillna(value=df.loan_amount.median()).plot.
kde(color='b', label='loan_amount_imputed', alpha=0.5, figsize=(9,7),
legend=True)
```

The following plot is displayed as the output:

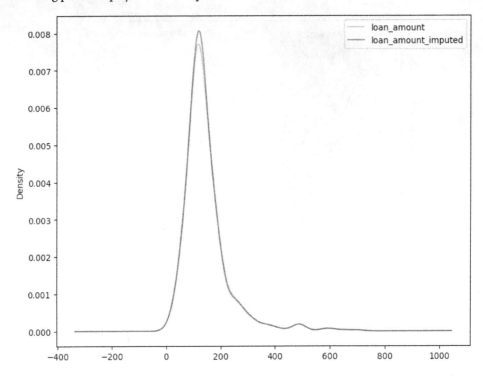

Figure 5.7 – Simple density plot imputation with the median

Next, we compare the standard deviation of the loan amount before and after imputation:

```
round(df_consistent.loan_amount.std(),2), round(df_consistent.loan_
amount.fillna(value=df_consistent.loan_amount.median()).std(),2)
(85.59, 84.11)
```

The preceding code shows how a simple imputation method can obscure the distribution of data. To counter these effects and preserve the distribution, we will use the random sample imputation method.

First, we extract all the rows where `loan_amount` is missing. Then, we compute the variables that are correlated with `loan_amount` and use those values to set a seed. This is because, if we use the same seed for all values, then the same random number will be generated and the method will behave similarly to arbitrary value imputation, which will be as ineffective as the simple imputation methods of mean and median.

The downside to random sample distribution is that covariance will be affected and we need a method that preserves the covariance as well.

First, we check which feature is highly correlated with `loan_amount`:

```
df_consistent[num_cols].corr()
                   applicant_income  coapplicant_income  loan_amount
applicant_income           1.000000           -0.116605     0.570909
coapplicant_income        -0.116605            1.000000     0.188619
loan_amount                0.570909            0.188619     1.000000
```

Here, we can see that `loan_amount` is highly correlated with `applicant_income`, so for this example, we use this variable to set the seed. First, we extract the indexes where `loan_amount` is missing. Then, we use the `applicant_income` value at the missing location and use this value to set the seed. Next, we use this seed to generate a random value from `loan_amount` to impute the missing row. We use this approach to impute all the missing data for `loan_amount`:

```
observation = df_consistent[df_consistent.loan_amount.isnull()]
imputed_values = []
for idx in observation.index:
    seed = int(observation.loc[idx,['applicant_income']])
    imputed_value = df_consistent['loan_amount'].dropna().sample(1,
random_state=seed)
    imputed_values.append(imputed_value)

df_consistent.loc[df_consistent['loan_amount'].isnull(),'loan_amount_
random_imputed']=imputed_values
df_consistent.loc[df['loan_amount'].isnull()==False,'loan_amount_
random_imputed']=df_consistent[df_consistent['loan_amount'].
isnull()==False]['loan_amount'].values
```

Next, we compare the distribution of `loan_amount` with the random sample imputed `loan_amount` and median imputed `loan_amount`:

```
df_consistent.loan_amount.plot.kde(color='orange', label='loan_
amount', legend=True, linewidth=2)
df_consistent.loan_amount_random_imputed.plot.kde(color='g',
label='loan_amount_random_imputed', legend=True, linewidth=2)
df_consistent.loan_amount.fillna(value=df_consistent.loan_amount.
median()).plot.kde(color='b', label='loan_amount_median_imputed',
linewidth=1, alpha=0.5, figsize=(9,7), legend=True)
<AxesSubplot: ylabel='Density'>
```

This will output the following plot:

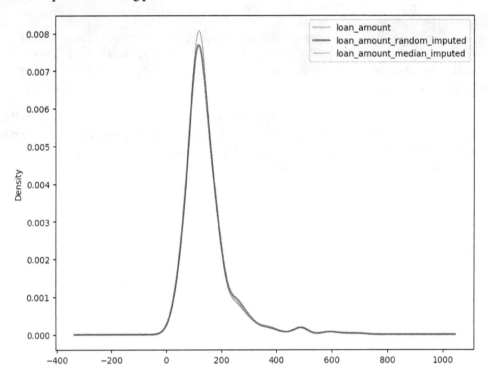

Figure 5.8 – Density plot showing random and median imputations

Now, we compare the standard deviation of the pre-imputed loan with the random sample imputed method and median imputed method:

```
round(df_consistent.loan_amount.std(),2), round(df_consistent.loan_
amount_random_imputed.std(),2), round(df_consistent.loan_amount.
fillna(value=df_consistent.loan_amount.median()).std(),2)
(85.59, 85.57, 84.11)
```

The random sample imputation method is much closer in distribution and standard deviation to the pre-imputed `loan_amount` method than the median imputed `loan_amount` method. Next, we check whether the random sample imputation method preserves the correlation with other variables compared to other methods:

```
df_consistent['loan_amount_median_imputed'] = df_consistent['loan_
amount'].fillna(value=df_consistent['loan_amount'].median())

df_consistent[['loan_amount', 'loan_amount_median_imputed','loan_
amount_random_imputed', 'applicant_income']].corr()
```

The resulting DataFrame is as follows:

	applicant_income	coapplicant_income	loan_amount
applicant_income	1.000000	-0.116605	0.570909
coapplicant_income	-0.116605	1.000000	0.188619
loan_amount	0.570909	0.188619	1.000000

Figure 5.9 – Correlation DataFrame

From this, it's evident that the random imputation method can preserve the distribution but may obscure the inter-relationships with other variables. We need a method that can preserve the distribution and also maintain the inter-relationships with other variables. We will use machine learning to help us achieve this. Before we move on to machine learning, we will first discuss the impact of simple imputation on categorical/binary variables. We impute the `credit_history` binary column with the most frequent value and compare the distribution before and after imputation:

```
df_consistent.credit_history.value_counts(normalize=True)
1.0     0.842199
0.0     0.157801
Name: credit_history, dtype: float64
df_consistent.credit_history.fillna(value=df_consistent.credit_
history.mode()[0]).value_counts(normalize=True)
1.0     0.855049
0.0     0.144951
Name: credit_history, dtype: float64
```

By imputing `credit_history` with the most frequent values, we have biased the data toward the `credit_history` status. As we discovered previously, the missingness of `credit_history` is not associated with any other variables, but it might be associated with the outcome.

The preceding examples show that if we utilize simple imputation methods then we may bias the data, and the distribution will be altered as well, whereas if we utilize random methods, the distribution will be preserved but the data relationships may change and data variance may increase. Hence, when data is MAR or MNAR, to achieve a balance between data bias and data variance, we can use a machine learning model.

To utilize machine learning for numerical imputation, we will leverage the nearest neighbors imputation method available in the `scikit-learn` library – `KNNImputer`. One issue with the imputer is that we can only pass a DataFrame to it, and not pass a list of columns. Hence, we will use the `SklearnTransformerWrapper` module, which is available as part of the `feature-engine` library, to pass a list of columns. Since KNN is a distance-based algorithm, to ensure that the model converges and one variable doesn't overpower the other variable, we must scale the data before using this algorithm.

Another technique to impute data is referred to as **Multiple Imputation by Chained Equations** (**MICE**). MICE works by imputing all the data with the mean, median, or mode. Then, regarding the variable that will be imputed, the initial imputed values are converted back into missing values. Then, using other variables as predictors, a machine learning model is used to predict missing values. After this, the next variable is imputed in a similar manner, where initially imputed values are converted back into missing values, and other variables, including the recently imputed variable, are used as predictors to impute the missing values. Once all the variables with missing values have been modeled and values have been imputed with predictions, the first round of imputation is completed. This procedure is repeated n number of times (ideally 10), and from round two, round one predictions are used to predict records that were initially missing.

The reason for using several rounds is that, initially, we are modeling the missing data using other variables that also have NA values, and the initial strategy of imputation uses suboptimal methods such as the mean, median, or mode, which may bias the predictions. As we continue to regress over multiple rounds, predictions will stabilize and become less biased.

One issue with MICE is that we have to choose which machine learning model to use for the task. We will implement MICE with the random forest algorithm, which in R language is referred to as [missForest].

In our implementation of MICE, we will refer to it as missForest since it will be a replica of how it is implemented in R (https://rpubs.com/lmorgan95/MissForest#:~:text=MissForest%20is%20a%20random%20forest,then%20predicts%20the%20missing%20part). To counter the effects of choosing an algorithm, we encourage practitioners to leverage automated machine learning, where for each imputation and iteration, a new algorithm will be chosen. The one disadvantage of this approach is that it's computationally intensive and time-intensive when utilized for big datasets.

First, we import the necessary packages:

```
from sklearn.impute import KNNImputer
from feature_engine.wrappers import SklearnTransformerWrapper
from sklearn.preprocessing import StandardScaler
```

Next, we extract the numerical columns by filtering any columns that may have more than 15 categories while filtering the id column and outcome, as well as filtering newly created variables using imputation methods:

```
num_cols = [cols for cols in df_consistent if df_consistent[cols].
nunique() > 15 and cols not in [id_col, target] and not cols.
endswith('imputed')]
```

Next, we create the DataFrame with numerical variables and visualize it:

```
df_num = df_consistent[num_cols].copy()
df_num.head()
```

```
    applicant_income  coapplicant_income  loan_amount
0               5849                 0.0          NaN
1               4583              1508.0        128.0
2               3000                 0.0         66.0
3               2583              2358.0        120.0
4               6000                 0.0        141.0
```

Next, we build a function that takes the scaler (standard scaler or any other scaler) and DataFrame and returns scaled data and the processed scaler. We must scale the dataset before applying the KNN imputer since a distance-based method requires data to be on the same scale. Once we have scaled the data, we apply the KNN imputer to impute the data, and then unscale the data using the processed scaler returned by the function. Once we've done this, we can compare the machine learning imputed data with the median and random imputation methods:

```python
def scale_data(df, scaler, columns):
    """Function to scale the data"""

    df_scaled = df.copy()
    if columns:
        df_scaled[columns] = scaler.fit_transform(df_scaled[columns])
    else:
        columns = [cols for cols in df_scaled]
        df_scaled[columns] = scaler.fit_transform(df_scaled[columns])

    return df_scaled, scaler
```

Next, we define the scaler and call the `scale_data` function:

```python
scaler = StandardScaler()
df_scaled, scaler = scale_data(df_num, scaler=scaler, columns=num_
cols)
```

Then, we apply the KNN imputer with a parameter of 10 neighbors to impute the data. We utilize the `weights='distance'` parameter so that more weightage will be given to the votes of the nearer neighbors compared to the ones that are further away when predicting the outcome.

First, we initialize the imputer:

```python
knn_imputer = SklearnTransformerWrapper(
    transformer = KNNImputer(n_neighbors=10, weights='distance'),
    variables = num_cols
)
```

Then, we apply the imputation:

```python
df_imputed = knn_imputer.fit_transform(df_scaled)
```

Next, we unscale the data by calling the `inverse_transform` method from the scaler object and overwrite the `df_imputed` DataFrame with unscaled values:

```
df_imputed = pd.DataFrame(columns=num_cols, data=scaler.inverse_
transform(df_imputed))
df_imputed.head()
     applicant_income  coapplicant_income  loan_amount
0               5849.0                 0.0   149.666345
1               4583.0              1508.0   128.000000
2               3000.0                 0.0    66.000000
3               2583.0              2358.0   120.000000
4               6000.0                 0.0   141.000000
```

Next, we compare the distribution of the pre-imputed `loan_amount` and compare it with the machine learning imputed method. Then, we check the correlation of the machine learning imputed method to the applicant's income and compare it with other imputed methods:

```
df_imputed['loan_amount'].plot.kde(color='orange', label='loan_amount_
knn_imputed',linewidth=2, legend=True)
df_consistent['loan_amount'].plot.kde(color='b', label='loan_amount',
legend=True, linewidth=2, figsize=(9,7), alpha=0.5)
```

The resulting plot is as follows:

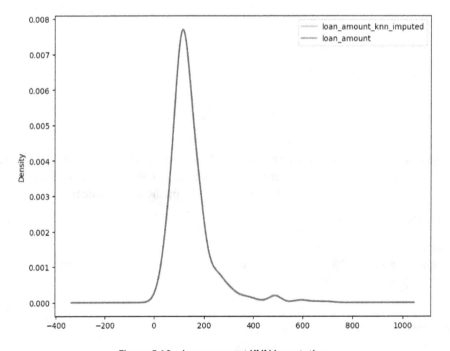

Figure 5.10 – Loan amount KNN imputation

Next, we compare the standard deviation of the pre-imputed loan amount with all the imputation methods:

```
round(df_consistent.loan_amount.std(),2), round(df_consistent.loan_
amount_random_imputed.std(),2), round(df_consistent.loan_amount_
median_imputed.std(),2), round(df_imputed.loan_amount.std(),2)
(85.59, 85.57, 84.11, 85.59)
```

Then, we will check if the correlation is maintained when `loan_amount` is imputed using machine learning:

```
df_consistent['loan_amount_knn_imputed'] = df_imputed.loan_amount
df_consistent[['loan_amount', 'loan_amount_median_imputed','loan_
amount_random_imputed', 'loan_amount_knn_imputed', 'applicant_
income']].corr()
```

	loan_amount	loan_amount_median_imputed	loan_amount_random_imputed	loan_amount_knn_imputed	applicant_income
loan_amount	1.000000	1.000000	1.000000	1.000000	0.570909
loan_amount_median_imputed	1.000000	1.000000	0.979281	0.981185	0.565181
loan_amount_random_imputed	1.000000	0.979281	1.000000	0.970700	0.561845
loan_amount_knn_imputed	1.000000	0.981185	0.970700	1.000000	0.578718
applicant_income	0.570909	0.565181	0.561845	0.578718	1.000000

Figure 5.11 – Correlation after loan_amount is imputed

The machine learning-imputed method has almost the same distribution as the original. However, the correlation is a bit higher with `applicant_income` compared to the pre-imputed `loan_amount`. We have now seen how to use out-of-the-box techniques to impute missing data. One advantage of this method is that it's simple to implement. However, the disadvantage is that we cannot choose another algorithm.

Hence, in the next step, we go one step further and build a MICE implementation with random forest. First, we convert categorical data into numerical data using one-hot encoding. Then, we impute missing categorical data with the MICE implementation with `RandomForestClassifier`.

Once the categorical data has been imputed, we use categorical and numerical data to impute numerical missing values by utilizing MICE with `RandomForestRegressor`.

To build the MICE implementation, we use `IterativeImputer`, which is available in scikit-learn, to help with 10 rounds of MICE. To leverage `IterativeImputer`, we must import `enable_iterative_imputer` from scikit-learn's experimental packages, as per the docs: https://scikit-learn.org/stable/modules/generated/sklearn.impute.IterativeImputer.html.

First, we import the necessary packages:

```
from sklearn.ensemble import ExtraTreesRegressor, ExtraTreesClassifier
from sklearn.experimental import enable_iterative_imputer
from sklearn.impute import IterativeImputer
from feature_engine.encoding import OneHotEncoder
```

Next, we extract the categorical columns that are string-encoded so that we can one-hot encode these:

```
ohe_cols = [cols for cols in cat_cols if df_consistent[cols].dtype ==
'object']
ohe_cols
['property_area']
```

Then, we one-hot encode the categorical columns:

```
df_ohe_encoded = df_consistent.copy()
ohe = OneHotEncoder(variables=ohe_cols)
df_ohe_encoded = ohe.fit_transform(df_ohe_encoded)
```

After that, we visualize the first five results of the one-hot encoded data:

```
df_ohe_encoded[[cols for cols in df_ohe_encoded if 'property_area' in
cols]].head()
```

	property_area_urban	property_area_rural	property_area_semiurban
0	1	0	0
1	0	1	0
2	1	0	0
3	1	0	0
4	1	0	0

Next, we extract the categorical variables that are binary encoded, including the data that has been one-hot encoded:

```
cat_cols = [cols for cols in df_ohe_encoded if df_ohe_encoded[cols].
nunique() <= 15 and cols not in [id_col, target]]
cat_cols
['gender',
 'married',
 'dependents',
 'education',
 'self_employed',
 'loan_amount_term',
 'credit_history',
 'property_area_urban',
 'property_area_rural',
 'property_area_semiurban']
```

Then, we build the MICE implementation with random forest to impute categorical data:

```
miss_forest_classifier = IterativeImputer(
    estimator=ExtraTreesClassifier(n_estimators=100,
                                    random_state=1,
                                    bootstrap=True,
                                    n_jobs=-1),
    max_iter=10,
    random_state=1,
    add_indicator=True,
    initial_strategy='median')

df_cat_imputed = miss_forest_classifier.fit_transform(df_ohe_
encoded[cat_cols])
```

Next, we extract the features from the imputation by converting the NumPy array into a DataFrame called `df_cat_imputed`:

```
df_cat_imputed = pd.DataFrame(
    columns=miss_forest_classifier.get_feature_names_out(),
    data=df_cat_imputed,
    index=df_ohe_encoded.index)
```

Let's ensure we don't have any new unexpected values being created by the classifier. To check this, we iterate over all the columns and print the unique values for each column:

```
for cols in cat_cols:
    print(cols)
    print(df_cat_imputed[cols].unique())
    print()
gender
[1. 0.]

married
[0. 1.]

dependents
[0. 1. 2. 3.]

education
[1. 0.]

self_employed
[0. 1.]
```

```
loan_amount_term
[360. 120. 240. 180.  60. 300. 480.  36.  84.  12.]

credit_history
[1. 0.]

property_area_urban
[1. 0.]

property_area_rural
[0. 1.]

property_area_semiurban
[0. 1.]
```

Now, we combine the categorical imputed data with numerical data. Then, we use all the data to impute numerical data:

```
num_cols = [cols for cols in df_consistent if cols not in df_cat_
imputed and cols not in [id_col, target] + ohe_cols
            and not cols.endswith("imputed")]

df_combined = pd.concat([df_consistent[num_cols], df_cat_imputed],
axis=1)
feature_cols = [cols for cols in df_combined]
feature_cols
['applicant_income',
 'coapplicant_income',
 'loan_amount',
 'gender',
 'married',
 'dependents',
 'education',
 'self_employed',
 'loan_amount_term',
 'credit_history',
 'property_area_urban',
 'property_area_rural',
 'property_area_semiurban',
 'missingindicator_gender',
 'missingindicator_married',
 'missingindicator_dependents',
 'missingindicator_self_employed',
```

```
'missingindicator_loan_amount_term',
'missingindicator_credit_history']
```

Next, we implement MICE imputation with random forest to impute numerical data:

```
miss_forest_regressor = IterativeImputer(
    estimator=ExtraTreesRegressor(n_estimators=100,
                                  random_state=1,
                                  bootstrap=True,
                                  n_jobs=-1),
    max_iter=10,
    random_state=1,
    add_indicator=True,
    initial_strategy='median')

df_imputed = miss_forest_regressor.fit_transform(df_combined[feature_
cols])
```

Now, we extract the features from the imputation by converting the NumPy array into a DataFrame:

```
df_imputed
df_imputed = pd.DataFrame(data=df_imputed,
                          columns=miss_forest_regressor.get_feature_
names_out(),
                          index=df_combined.index)
```

Then, we check whether all the columns have been imputed and there are no missing values:

```
df_imputed.isnull().sum()
applicant_income                      0
coapplicant_income                    0
loan_amount                           0
gender                                0
married                               0
dependents                            0
education                             0
self_employed                         0
loan_amount_term                      0
credit_history                        0
property_area_urban                   0
property_area_rural                   0
property_area_semiurban               0
missingindicator_gender               0
missingindicator_married              0
missingindicator_dependents           0
```

```
missingindicator_self_employed        0
missingindicator_loan_amount_term      0
missingindicator_credit_history        0
missingindicator_loan_amount           0
dtype: int64
```

Next, we compare the distribution of the pre-imputed `loan_amount` and compare it with the MICE imputed method. We then check the correlation of the MICE imputed method with the applicant's income and compare it with other imputed methods:

```
df_imputed['loan_amount'].plot.kde(color='orange', label='loan_amount_
miss_forest_imputed',linewidth=2, legend=True)

df_consistent['loan_amount'].plot.kde(color='b', label='loan_amount',
legend=True, linewidth=2, figsize=(9,7), alpha=0.5)

<AxesSubplot: ylabel='Density'>
```

The output is as follows:

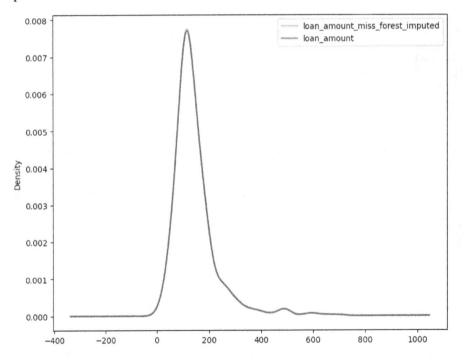

Figure 5.12 – loan_amount_miss_forest_imputed

Next, we compare the standard deviation of the pre-imputed loan amount with all the imputation methods, including the MICE imputation method:

```
round(df_consistent.loan_amount.std(),2), round(df_consistent.loan_
amount_random_imputed.std(),2), round(df_consistent.loan_amount_
median_imputed.std(),2), round(df_imputed.loan_amount.std(),2)
(85.59, 85.57, 84.11, 85.41)
```

Then, we check if the correlation is maintained when `loan_amount` is imputed using the MICE imputation method compared to other methods:

```
df_consistent['loan_amount_miss_forest_imputed'] = df_imputed.loan_
amount
df_consistent[['loan_amount', 'loan_amount_median_imputed','loan_
amount_random_imputed', 'loan_amount_miss_forest_imputed', 'applicant_
income']].corr()
```

The output DataFrame is as follows:

	loan_amount	loan_amount_median_imputed	loan_amount_random_imputed	loan_amount_miss_forest_imputed	applicant_income
loan_amount	1.000000	1.000000	1.000000	1.000000	0.570909
loan_amount_median_imputed	1.000000	1.000000	0.979281	0.982379	0.565181
loan_amount_random_imputed	1.000000	0.979281	1.000000	0.972310	0.561845
₃an_amount_miss_forest_imputed	1.000000	0.982379	0.972310	1.000000	0.579011
applicant_income	0.570909	0.565181	0.561845	0.579011	1.000000

Figure 5.13 – Correlation after imputing loan_amount using the MICE imputation method

The standard deviation is slightly below the random imputation but higher than the median imputed method. As we can see, the correlation with `applicant_income` hasn't improved compared to the random imputation method or median imputation method. Hence, to test whether MICE with random forest is a better implementation for this use case, we can compare the evaluation metric of the machine learning model when MICE is utilized versus when median imputation is utilized.

But before we do that, we want machine learning practitioners to explore **automated machine learning (AutoML)** with the MICE imputation framework. Machine learning can be a tedious process that consists of trial and error, hence why AutoML frameworks are getting quite popular when it comes to reducing human time. These frameworks automate feature engineering, cross-validation, model selection, and model tuning. One issue with the current implementation of MICE is that we have to choose which machine learning model to use for the task. What if we wanted to trial multiple algorithms to see which provided the best prediction for the imputation task, and while doing that wanted to ensure the prediction was generalizable and the model was not overfitted or underfitted? We can imagine the complexity. To counter this, we'll combine AutoML with MICE.

One advantage of this approach is that at each iteration, a new model will be picked by AutoML, thus freeing the machine learning practitioner from tedious tasks. However, the disadvantage of this approach is that when the data increases in size, a lot more resources will be needed, which may not be viable. Another disadvantage with some open source AutoML frameworks is that on some operating systems, full functionality is error-prone. For instance, on Mac computers, both TPOT and AutoSklearn frameworks give errors when parallel processing is used. Hence, we will let you explore your own flavor of AutoML with MICE.

Next, we will implement a scikit-learn pipeline that will include the MICE implementation with random forest. Then, we train the decision tree model with cross-validation and evaluate the model using accuracy and ROC. Once we've done this, we create another pipeline, which will use simple imputation methods, and compare the evaluation results. Finally, we explore techniques for further improving the data to enhance model performance.

We'll be converting these steps into a scikit-learn pipeline since by using a pipeline, we can define the sequence of steps and also save these steps as a pickle object. By utilizing this practice, we maintain machine learning system best practices and can ensure reliability and reproducibility without replicating the code in the inference environment.

First, let's drop all the newly created columns in the df_consistent DataFrame that end with _ imputed:

```
df_consistent.drop([cols for cols in df_consistent if cols.
endswith('imputed')], axis=1, inplace=True)
```

Next, we import all the necessary packages and modules to help split the data into train and test sets, evaluate the performance of the model, and create a machine learning pipeline:

```
from sklearn.model_selection import train_test_split, GridSearchCV
from sklearn.metrics import roc_auc_score, accuracy_score, confusion_
matrix, ConfusionMatrixDisplay
from typing import List
from sklearn.tree import DecisionTreeClassifier
from sklearn.preprocessing import StandardScaler
from sklearn.pipeline import Pipeline
from sklearn.compose import ColumnTransformer
from sklearn.impute import SimpleImputer
```

Now, we extract the features for the model and split the data into train and test sets, where 10% of the data is reserved for testing:

```
feature_cols = [cols for cols in df_consistent if cols not in [target,
id_col]]
X_train, X_test, y_train, y_test = train_test_split(df_
consistent[feature_cols],
                                                     df_
consistent[target].map({'Y':1, 'N':0}),
```

```
                                            test_size=0.1,
                                            random_state=1,
                                            stratify=df_
consistent[target].map({'Y':1, 'N':0}))
feature_cols
['gender',
 'married',
 'dependents',
 'education',
 'self_employed',
 'applicant_income',
 'coapplicant_income',
 'loan_amount',
 'loan_amount_term',
 'credit_history',
 'property_area']
```

Next, we extract the categorical data and numerical data into separate lists so that we can use these to set the pipeline for each type of data:

```
cat_cols = [cols for cols in X_train if X_train[cols].nunique() <= 15]
num_cols = [cols for cols in X_train if cols not in cat_cols]
```

Now, we create a function that will return the pipeline for categorical data. First, the pipeline one-hot encodes the list of columns in the `ohe_cols` variable, which includes `property_area`. The pipeline then imputes the columns with missing data using the MICE implementation with random forest. The function will return the transformer so that when we pass the categorical data, while the transformer one-hot encodes the data and then imputes the missing data. The transformer will be run against the training data first so that it learns about the data and saves all the metadata for running the same steps with new data. The transformer can then be used to transform the test data:

```
def miss_forest_categorical_transformer():
    """Function to define categorical pipeline"""

    cat_transformer = Pipeline(
        steps=[
            ("one_hot_encoding",
             OneHotEncoder(variables=ohe_cols)
            ),

            ("miss_forest_classifier",
             IterativeImputer(
                 estimator=ExtraTreesClassifier(
                     n_estimators=100,
```

```
                                  random_state=1,
                                  bootstrap=True,
                                  n_jobs=-1),
                    max_iter=10,
                    random_state=1,
                    initial_strategy='median',
                    add_indicator=True)
            )
        ]
    )

    return cat_transformer
```

Next, we create a function that returns the pipeline transformer to impute numerical missing data with the MICE implementation. Similarly to the categorical transformer, the numerical transformer will be trained against the training data and then applied to the test data to impute missing values in the train and test data:

```
def miss_forest_numerical_transformer():
    """Function to define numerical pipeline"""

    num_transformer = Pipeline(
        steps=[
            ("miss_forest",
             IterativeImputer(
                estimator=ExtraTreesRegressor(n_estimators=100,
                                              random_state=1,
                                              bootstrap=True,
                                              n_jobs=-1),
                max_iter=10,
                random_state=1,
                initial_strategy='median',
                add_indicator=True)
            )
        ]
    )

    return num_transformer
```

Then, we initialize the categorical and numerical transformers, and then transform training and test data. The transformed categorical data is combined with numerical data before the numerical data is transformed. The output of this is imputed train and test DataFrames:

```
cat_transformer = miss_forest_categorical_transformer()
num_transformer = miss_forest_numerical_transformer()


X_train_cat_imputed = cat_transformer.fit_transform(X_train[cat_cols])
X_test_cat_imputed = cat_transformer.transform(X_test[cat_cols])

X_train_cat_imputed_df = pd.DataFrame(data=X_train_cat_imputed,
                                      columns=cat_transformer.get_
feature_names_out(),
                                      index=X_train.index)

X_test_cat_imputed_df = pd.DataFrame(data=X_test_cat_imputed,
                                     columns=cat_transformer.get_
feature_names_out(),
                                     index=X_test.index)

X_train_cat_imputed_df = pd.concat([X_train_cat_imputed_df, X_
train[num_cols]], axis=1)
X_test_cat_imputed_df = pd.concat([X_test_cat_imputed_df, X_test[num_
cols]], axis=1)


X_train_imputed = num_transformer.fit_transform(X_train_cat_imputed_
df)
X_test_imputed = num_transformer.transform(X_test_cat_imputed_df)

X_train_transformed = pd.DataFrame(data=X_train_imputed,
                                   columns=num_transformer.get_
feature_names_out(),
                                   index=X_train.index)

X_test_transformed = pd.DataFrame(data=X_test_imputed,
                                  columns=num_transformer.get_feature_
names_out(),
                                  index=X_test.index)
```

Before passing the complete datasets to the machine learning model, we check if both the train and test labels have similar loan approval rates:

```
y_train.mean(), y_test.mean()
(0.6865942028985508, 0.6935483870967742)
```

Because the classes are slightly imbalanced, we can use the `class_weight='balanced'` option since this option uses the values of y to automatically adjust weights inversely proportional to class frequencies in the input data when training the algorithm. The objective of the problem is to identify better than a human being who is likely to get a loan. Since the majority class is trained on people who received a loan, the model will be biased toward giving someone a loan. By using `class_weight='balanced'`, the algorithm will put more emphasis on class label 0 since it's a minority class.

We define grid search for the decision tree classifier to perform cross-validation, to ensure the model is generalizable:

```
d_param_grid = {
    'max_features': [None, 'sqrt', 'log2'],
    'max_depth' : [4,5,6,7,8,10,20],
    'min_samples_leaf' : [1,3,5,8,10,12,15],
    'min_samples_split': [2,6,10,16,20,24,30],
    'criterion' : ['gini', 'entropy'],
    'random_state' : [1],
    'class_weight' : ['balanced']
}
d_clf = DecisionTreeClassifier()
```

Next, we create a custom function that will take in training data, testing data, the classifier, and grid search parameters. The function performs 10K cross-validation to find the best hyperparameters and trains the model on the best parameters. The function then returns the model, predictions, training and test accuracies, and ROC-AUC score:

```
def train_custom_classifier(X_train, y_train, X_test, y_test, clf,
params):
    """Function to train the decision tree classifier and return some
metrics"""

    d_clf_cv = GridSearchCV(estimator=d_clf, param_grid=d_param_grid,
cv=10, scoring='roc_auc')
    d_clf_cv.fit(X_train_transformed, y_train)

    print("Decision tree optimised")


    d_best_params = d_clf_cv.best_params_

    print(f"Getting the best params which are {d_best_params}")

    model = DecisionTreeClassifier(**d_best_params)
    model.fit(X_train_transformed, y_train)
```

```
    training_predictions_prob = model.predict_proba(X_train_
transformed)
    testing_predictions_prob = model.predict_proba(X_test_transformed)

    training_predictions = model.predict(X_train_transformed)
    testing_predictions = model.predict(X_test_transformed)

    training_roc_auc = roc_auc_score(y_train, training_predictions_
prob[:,1])
    testing_roc_auc = roc_auc_score(y_test, testing_predictions_
prob[:,1])

    training_acc = accuracy_score(y_train, training_predictions)
    testing_acc = accuracy_score(y_test, testing_predictions)

    print(f"Training roc is {training_roc_auc}, and testing roc is
{testing_roc_auc} \n \
            training accuracy is {training_acc}, testing_acc as
{testing_acc}")

    return model, testing_predictions, training_roc_auc, testing_roc_
auc, training_acc, testing_acc
```

Next, we run the custom classifier and calculate model performance:

```
model, test_predictions, train_roc, test_roc, train_acc, test_acc  =
train_custom_classifier(
    X_train=X_train_transformed,
    y_train=y_train,
    X_test=X_test_transformed,
    y_test=y_test,
    clf=d_clf,
    params=d_param_grid

)
Decision tree optimised
Getting the best params which are {'class_weight': 'balanced',
'criterion': 'entropy', 'max_depth': 8, 'max_features': None, 'min_
samples_leaf': 1, 'min_samples_split': 30, 'random_state': 1}
Training roc is 0.8763326063416048, and testing roc is
0.7858017135862914
            training accuracy is 0.8152173913043478, testing_acc as
0.7903225806451613
```

The test accuracy is just under 80%. Let's see where the model is performing poorly by observing the confusion matrix:

```
cm = confusion_matrix(y_test, test_predictions, labels=model.classes_,
normalize='true')
disp = ConfusionMatrixDisplay(confusion_matrix=cm, display_
labels=model.classes_)
disp.plot()
```

This outputs the following confusion matrix:

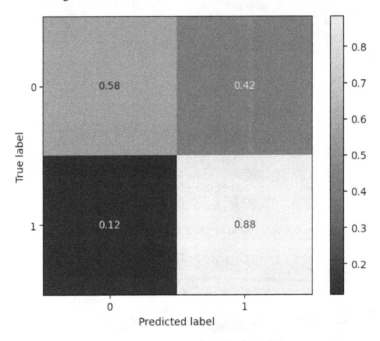

Figure 5.14 – Confusion matrix

We have now applied machine learning pipelines with MICE imputation to create a machine learning model. To demonstrate that MICE imputation is a better technique than the simple imputation technique, we will recreate the machine learning pipelines with simple imputation methods and evaluate model performance.

Once we have created the pipeline steps, we transform the train and test data before passing it to the decision tree classifier and custom classifier function to measure model performance:

```
cat_transformer = Pipeline(
    steps=[
        ("one_hot_encoding",
```

```
                OneHotEncoder(variables=ohe_cols)
            )
        ]
)

impute_transformer = Pipeline(
    steps=[
        ("simple_imputer",
        SimpleImputer(strategy='median',
                        add_indicator=True)
        )
    ]
)

X_train_ohe = cat_transformer.fit_transform(X_train)
X_test_ohe = cat_transformer.transform(X_test)


X_train_imputed = impute_transformer.fit_transform(X_train_ohe)
X_test_imputed = impute_transformer.transform(X_test_ohe)


X_train_transformed = pd.DataFrame(data=X_train_imputed,
                                    columns=impute_transformer.get_
feature_names_out(),
                                    index=X_train.index)

X_test_transformed = pd.DataFrame(data=X_test_imputed,
                                    columns=impute_transformer.get_
feature_names_out(),
                                    index=X_test.index)
```

Next, we run the custom classifier and extract model performance:

```
model, test_predictions, train_roc, test_roc, train_acc, test_acc =
train_custom_classifier(
    X_train=X_train_transformed,
    y_train=y_train,
    X_test=X_test_transformed,
    y_test=y_test,
    clf=d_clf,
    params=d_param_grid
)
```

The test accuracy has dropped under 67%, which is a 12% reduction, and the ROC-AUC has dropped by 6%. Next, we review the confusion matrix:

```
cm = confusion_matrix(y_test, test_predictions, labels=model.classes_,
normalize='true')
disp = ConfusionMatrixDisplay(confusion_matrix=cm, display_
labels=model.classes_)
disp.plot()
```

Here's the output:

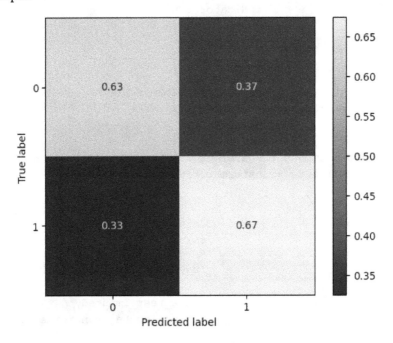

Figure 5.15 – Confusion matrix

The accuracy of the true positive class has dropped from 88% to 67%, whereas the accuracy of the negative class has increased from 58% to 63%. By using basic imputation techniques, we can conclude that the model is more likely to be biased and the model performance may be less accurate.

In data-centric machine learning, the goal is to improve on the data and tune it, rather than improving on the algorithm and tuning the model. But how can we identify whether a dataset contains poorly labeled data, missing features, or another data-related issue? We will cover how to identify if data is poorly labeled and apply techniques to improve on mislabeled data in *Chapter 6, Techniques for Programmatic Labeling in Machine Learning*.

To find out if more features are needed or more data is needed, we utilize a technique called error analysis. In machine learning, error analysis is utilized to identify and diagnose erroneous predictions by focusing on the pockets of data where the model performed well and poorly. Although the overall performance of the model might be 79%, this performance may not be uniform across all pockets of the data, and these highs and lows could be due to inputs present in some pockets and absent in other pockets of data.

To identify data issues, we will start training the model with 10% of the data, and with each iteration add 10%. Then, we plot the training ROC and test the ROC concerning the increase in the size of the data. If the plot seems to converge and indicate an increase in the size of the data, this will lead to an improvement in the test ROC, at which point we will generate synthetic data to increase the data's size. This technique will be covered in *Chapter 7, Using Synthetic Data in Data-Centric Machine Learning*.

If the plot doesn't seem to converge and indicates an increase in data, it will have a minimal impact on improving test ROC. In this case, we can observe which data points the model performed poorly on, and may utilize feature engineering to generate new columns. Although feature engineering can be an iterative approach, for the scope of this chapter, we cover adding a feature or two.

To run error analysis, first, we create data cutoff points from 0.1 to 1.0, where 0.1 means 10% of the training data and 1.0 means 100% of the training data:

```
data_cutoff_points = np.linspace(start=0.1, stop=1, num=10)
data_cutoff_points
array([0.1, 0.2, 0.3, 0.4, 0.5, 0.6, 0.7, 0.8, 0.9, 1. ])
```

Next, we create an empty list called `scores` and run data preprocessing, model training, and evaluation with each cutoff of data. If the cutoff is < 1.0, we subset the training data; otherwise, we pass all the data for training. At the end of each iteration, we save the cutoff, train, and test evaluation metrics in `scores` by appending the metrics to the `scores` list:

```
scores = []
for cutoff in data_cutoff_points:
    if cutoff < 1.0:
        X_train_subset, X_train_rem, y_train_subset, y_train_rem =
train_test_split(X_train,
                    y_train,
                        random_state=1,
                            train_size=cutoff,
                        stratify=y_train)
    else:
        X_train_subset = X_train.copy()
        y_train_subset = y_train.copy()

    print(f"Model will be trained on {X_train_subset.shape[0]} rows
out of {X_train.shape[0]}")
```

```
    cat_transformer = miss_forest_categorical_transformer()
    num_transformer = miss_forest_numerical_transformer()


    X_train_cat_imputed = cat_transformer.fit_transform(X_train_
subset[cat_cols])
    X_test_cat_imputed = cat_transformer.transform(X_test[cat_cols])


    X_train_cat_imputed_df = pd.DataFrame(data=X_train_cat_imputed,
                                          columns=cat_transformer.
get_feature_names_out(),
                                          index=X_train_subset.index)


    X_test_cat_imputed_df = pd.DataFrame(data=X_test_cat_imputed,
                                         columns=cat_transformer.get_
feature_names_out(),
                                         index=X_test.index)


    X_train_cat_imputed_df = pd.concat([X_train_cat_imputed_df, X_
train_subset[num_cols]], axis=1)
    X_test_cat_imputed_df = pd.concat([X_test_cat_imputed_df, X_
test[num_cols]], axis=1)


    X_train_imputed = num_transformer.fit_transform(X_train_cat_
imputed_df)
    X_test_imputed = num_transformer.transform(X_test_cat_imputed_df)


    X_train_transformed = pd.DataFrame(data=X_train_imputed,
                                       columns=num_transformer.get_
feature_names_out(),
                                       index=X_train_subset.index)


    X_test_transformed = pd.DataFrame(data=X_test_imputed,
                                      columns=num_transformer.get_
feature_names_out(),
                                      index=X_test.index)


    model, test_predictions, train_roc, test_roc, train_acc, test_acc
= train_custom_classifier(
        X_train=X_train_transformed,
        y_train=y_train_subset,
        X_test=X_test_transformed,
        y_test=y_test,
        clf=d_clf,
```

```
        params=d_param_grid)

    scores.append((cutoff, train_roc, test_roc, train_acc, test_acc))
Model will be trained on 55 rows out of 552
Training roc is 0.9094427244582044, and testing roc is
0.5917992656058751
            training accuracy is 0.7454545454545455, testing_acc as
0.5806451612903226
Model will be trained on 110 rows out of 552
Training roc is 0.901702786377709, and testing roc is
0.7552019583843328
            training accuracy is 0.7272727272727273, testing_acc as
0.6290322580645161
Model will be trained on 165 rows out of 552
Training roc is 0.8986555479918311, and testing roc is
0.7099143206854346
            training accuracy is 0.7696969696969697, testing_acc as
0.5967741935483871
Model will be trained on 220 rows out of 552
Training roc is 0.8207601497264613, and testing roc is
0.8084455324357405
            training accuracy is 0.8318181818181818, testing_acc as
0.8064516129032258
Model will be trained on 276 rows out of 552
Training roc is 0.8728942407103326, and testing roc is
0.7906976744186047
            training accuracy is 0.822463768115942, testing_acc as
0.7419354838709677
Model will be trained on 331 rows out of 552
Training roc is 0.9344501863774991, and testing roc is
0.7753977968176254
            training accuracy is 0.8368580060422961, testing_acc as
0.7419354838709677
Model will be trained on 386 rows out of 552


Training roc is 0.8977545610478715, and testing roc is
0.7184822521419829
            training accuracy is 0.7849740932642487, testing_acc as
0.6612903225806451
Model will be trained on 441 rows out of 552
Training roc is 0.8954656335198737, and testing roc is
0.7429620563035496
            training accuracy is 0.81859410430839, testing_acc as
0.7258064516129032
Model will be trained on 496 rows out of 552
Training roc is 0.9102355500898685, and testing roc is
0.7441860465116278
```

```
            training accuracy is 0.8266129032258065, testing_acc as
0.7258064516129032
Model will be trained on 552 rows out of 552
Training roc is 0.8763326063416048, and testing roc is
0.7858017135862914
            training accuracy is 0.8152173913043478, testing_acc as
0.7903225806451613
```

Next, we create a DataFrame from the `scores` list and pass the relevant column names:

```
df = pd.DataFrame(data=scores, columns=['data_size', 'training_roc',
'testing_roc', "training_acc", "testing_acc"])
```

Then, we plot the train and test ROC against each cutoff:

```
plt.plot(df.data_size, df.training_roc, label='training_roc')
plt.plot(df.data_size, df.testing_roc, label='testing_roc')
plt.xlabel("Data Size")
plt.ylabel("ROC")
plt.title("Error Analysis")
plt.legend()
```

This will output the following plot:

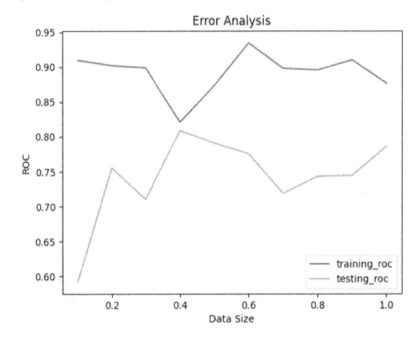

Figure 5.15 – Error analysis train and test ROC

Next, plot the train and test accuracy against each cutoff:

```
plt.plot(df.data_size, df.training_acc, label='training_acc')
plt.plot(df.data_size, df.testing_acc, label='testing_acc')
plt.xlabel("Data Size")
plt.ylabel("Accuracy")
plt.title("Error Analysis")
plt.legend()
```

This will output the following plot:

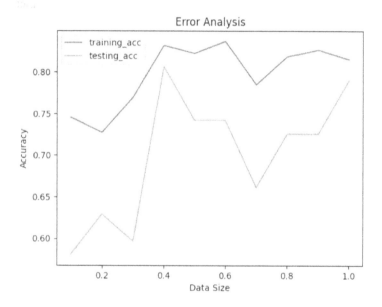

Figure 5.17 – Error analysis train and test accuracy

Both the test ROC and test accuracy seem to show signs of convergence with the train ROC and train accuracy, which indicates that model performance may be boosted if more data points were made available. This is why we will generate synthetic data (data that mimics the real data) in the next chapter and retrain the model with added data to get better model performance.

As we learned in the previous chapters, one of the principles of data-centric machine learning is keeping humans in the loop. Let's imagine we spoke to the domain experts and they mentioned that one of the key determinants for someone getting a loan is the income-to-debt ratio – that is, the total income divided by the loan amount. This determines if someone will be able to pay back the loan or not. An application with a lower income-to-loan ratio is more likely to be rejected. In the dataset, there are two income variables – applicant income and co-applicant income. Also, the loan amount is represented in thousand figures – that is, a loan amount in the data loan amount of 66 represents 66,000. To create this ratio, we will multiply the loan amount by 1,000 and then combine the income of the applicant and co-applicant. Once we've done this, we will divide the combined income by the loan amount to get the income-to-loan ratio. The domain experts also mentioned that **equated monthly installments (EMIs)** can also determine a candidate's capability to pay the loan. The lower the EMI, the more likely a loan will be accepted, whereas the higher the EMI, the more likely a loan will be rejected. To calculate this without the interest rate, we can use the loan term and loan amount to get an approximate EMI amount for each month.

For the income-to-loan ratio, we will create a custom transformer for multiplying the loan amount by 1,000 so that we can use it in the pipeline.

This transformer is a Python class that we can use to overload the fit and transform functions required by the pipeline. This class will inherit from the `BaseEstimator` and `TransformerMixin` classes, both of which can be found in the `sklearn.base` module. The class will be used to implement the fit and transform methods. These methods should contain X and y parameters, and the transform method should return a pandas DataFrame to ensure compatibility with the scikit-learn pipeline.

To create a full income column, we leverage the `feature_engine` library since it is already compatible with the scikit-learn pipeline and has methods to apply mathematical operations relative to other variables. First, we sum the income variables. The output of that transformation will be divided by the `loan_amount` variable to create the income-to-loan ratio.

To create the EMI, we leverage the `feature_engine` library and divide `loan_amount` with `loan_amount_term`. Once we have created these features, we remove the two income variables since we already created a combination of the two. For this step, we use the `DropFeatures` class from the `feature_engine` library. All these feature engineering steps will be combined in a new pipeline called `feature-transformer` and will be applied post-data imputation.

We believe that by adding these extra features, the model performance of the decision tree algorithm will improve. Let's run the algorithm post-feature engineering and evaluate the results.

First, we create custom variables that will take in a list of variables for the feature engineering steps:

```python
income_variables = ['applicant_income', 'coapplicant_income']
loan_variable = ['loan_amount']
loan_term_variable = ['loan_amount_term']
```

Next, we import relevant packages from `feature_engine` to perform the feature engineering steps and import the `BaseEstimator` and `TransformerMixin` classes:

```python
from feature_engine.creation.math_features import MathFeatures
from feature_engine.creation.relative_features import RelativeFeatures
from sklearn.base import BaseEstimator, TransformerMixin
from feature_engine.selection import DropFeatures
```

Then, we create a custom transformer that will take in variable names and a value that will be multiplied by each variable. By default, each variable will be multiplied by 1:

```
class MultiplyColumns(BaseEstimator, TransformerMixin):
    """Custom pipeline class to multiply columns passed in a DataFrame
with a value"""

    def __init__(self, multiply_by=1, variables=None):
        self.multiply_by = multiply_by
        self.variables = variables

    def fit(self, X, y=None):
        return self

    def transform(self, X, y=None):
        if self.variables:
            X[self.variables] = X[self.variables] * self.multiply_by
        return X
```

Next, we call the `missForest` categorical and numerical transformers we created previously. Once we've done this, we create a feature transformer pipeline that multiplies `loan_amount` by 1,000 by leveraging the custom transformer we created previously. The new pipeline then adds income variables to create one income variable, the income-to-loan ratio, and the EMI features. Finally, the pipeline drops the two income variables since the new income variable will be created. By using the transformer pipelines, the train and test data will be transformed, and new features will be created. The output of this step will be fully transformed into train and test data so that it can be passed to the custom classifier:

```
cat_transformer = miss_forest_categorical_transformer()
num_transformer = miss_forest_numerical_transformer()


feature_transformer = Pipeline(
    steps=[
        ("multiply_by_thousand",
         MultiplyColumns(
             multiply_by=1000,
             variables=loan_variable
         )
        ),
        ("add_columns",
         MathFeatures(
             variables=income_variables,
             func='sum'
```

```
        )
      ),
      ("income_to_loan_ratio",
       RelativeFeatures(variables=[f"sum_{income_variables[0]}_
{income_variables[1]}"],
                        reference=loan_variable,
                        func=["div"]
                        )
      ),
      ("emi",
       RelativeFeatures(variables=loan_variable,
                        reference=loan_term_variable,
                        func=["div"])
      ),
      ("drop_features",
       DropFeatures(features_to_drop=income_variables
        ))
    ]
)
```

Next, we create the categorical transformers for imputation:

```
X_train_cat_imputed = cat_transformer.fit_transform(X_train[cat_cols])
X_test_cat_imputed = cat_transformer.transform(X_test[cat_cols])

X_train_cat_imputed_df = pd.DataFrame(data=X_train_cat_imputed,
                                      columns=cat_transformer.get_
feature_names_out(),
                                      index=X_train.index)

X_test_cat_imputed_df = pd.DataFrame(data=X_test_cat_imputed,
                                     columns=cat_transformer.get_
feature_names_out(),
                                     index=X_test.index)

X_train_cat_imputed_df = pd.concat([X_train_cat_imputed_df, X_
train[num_cols]], axis=1)
X_test_cat_imputed_df = pd.concat([X_test_cat_imputed_df, X_test[num_
cols]], axis=1)
```

Then, we add the numerical imputation and complete the imputation steps:

```
X_train_imputed = num_transformer.fit_transform(X_train_cat_imputed_
df)
X_test_imputed = num_transformer.transform(X_test_cat_imputed_df)

X_train_imputed_df = pd.DataFrame(data=X_train_imputed,
                                  columns=num_transformer.get_
feature_names_out(),
                                  index=X_train.index)

X_test_imputed_df = pd.DataFrame(data=X_test_imputed,
                                 columns=num_transformer.get_feature_
names_out(),
                                 index=X_test.index)
```

Next, we transform the imputed data using the feature imputation pipeline we created previously:

```
X_train_transformed = feature_transformer.fit_transform(X_train_
imputed_df)
X_test_transformed = feature_transformer.transform(X_test_imputed_df)
```

At this point, we call the custom classifier function to evaluate model performance with added feature engineering steps:

```
model, test_predictions, train_roc, test_roc, train_acc, test_acc =
train_custom_classifier(
    X_train=X_train_transformed,
    y_train=y_train,
    X_test=X_test_transformed,
    y_test=y_test,
    clf=d_clf,
    params=d_param_grid)
Training roc is 0.8465996614150411, and testing roc is
0.8188494492044063
            training accuracy is 0.8206521739130435, testing_acc as
0.8225806451612904
```

Next, we call the confusion matrix:

```
cm = confusion_matrix(y_test, test_predictions, labels=model.classes_,
normalize='true')
disp = ConfusionMatrixDisplay(confusion_matrix=cm, display_
labels=model.classes_)
disp.plot()
```

The resulting confusion matrix is as follows:

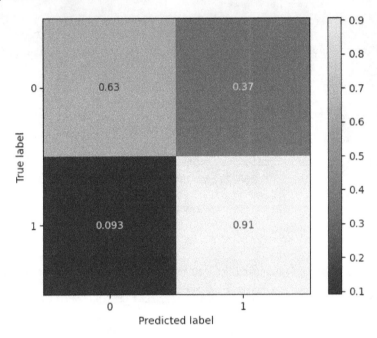

Figure 5.18 – Confusion matrix when using custom feature engineering

Our test accuracy has increased from 79% to 82% and the ROC has been boosted from 78.5% to 81.8%. The preceding confusion matrix shows that the accuracy of the positive class has been boosted from 88% to 91%, while the accuracy of the negative class has been boosted from 58% to 63%.

With this, we have demonstrated that by using a data-centric approach, we can iterate over the data rather than iterate over multiple algorithms, and manage to improve model performance. We will explore how to add some synthetic data and boost model performance even further in the next chapter.

Ensuring that the data is valid

So far, we have ensured that our data is consistent, unique, and complete. But do we know if the data we have is valid? Do the data labels conform to the rules? For example, what if the property area in the dataset didn't conform to the rules and `semi_urban` is invalid? What if one or a couple of annotators believed some suburbs are neither urban nor rural, and they violated the rules and entered `semi_urban`? To measure validity, we may need to look at business rules and check the percentage of data that conforms to these business rules. Let's assume that `semi_urban` is an invalid value. In Python, we could check the percentage of invalid labels and then reach out to annotators to correct the data. We could also achieve this by using the data that was used to generate the label. If we had the `suburb_name` to `property_area` data mapping, and `suburb_name` was available in the dataset, then we could leverage the mapping and catch invalid values, as well as encoding labels programmatically. Building business rules in the system so that upcoming data is automatically encoded is referred to as programmatic labeling. We will dive into programmatic labeling in the upcoming chapters, where we explore techniques to make data consistent and valid at the time of data capture, so that when the data comes in, it's already pristine and some of the data cleaning process will be redundant.

First, we make a fake dataset that consists of 10 rows and write a business rule. It will contain an `id` column with values from 1 to 10, a `population` column with 10 random values between 1,000 and 100,000, and a `property_area` column with four values set to `urban`, five values set to `semi_urban`, and one value set to `rural`:

```
np.random.seed(1)
data = {
    "id": np.linspace(start=1, stop=10, num=10, dtype=int),
    "population" : np.random.randint(low=1000, high=100000, size=10),
    "property_area": ["urban"]*4 + ["semi_urban"]*5 + ["rural"]*1
}

df = pd.DataFrame(data=data)
```

Next, we print the first five rows:

```
df.head()
   id  population property_area
0   1       99539         urban
1   2       78708         urban
2   3        6192         urban
3   4       99047         urban
4   5       51057    semi_urban
```

Imagine that the business rule says a suburb or `property_area` is classified as urban when the population is more than 20,000; otherwise, it's classified as rural. In this case, the validation rule should check that `property_area` only contains urban or rural values.

A simple way to check this in Python is to use the `value_counts()` method alongside the `normalize=True` parameter. The output of this will show that 50% of the data is invalid:

```
df.property_area.value_counts(normalize=True)
semi_urban    0.5
urban         0.4
rural         0.1
Name: property_area, dtype: float64
```

Next, we can run the check against each row and flag when the value is in the expected list of values, as well as when the value is not in the expected set of values:

```
df.property_area.isin(['rural', 'urban']) == False
0     False
1     False
2     False
3     False
4      True
5      True
6      True
7      True
8      True
9     False
Name: property_area, dtype: bool
```

Now, we sum the rows where the data validation rule was breached and divide the number of invalid rows by the total rows to provide a metric – that is, the percentage of invalid labels:

```
sum(df.property_area.isin(['rural', 'urban']) == False) / df.shape[0]
0.5
```

Invalid data must be communicated to the source data providers and must be cleaned; otherwise, this data can creep in and the machine learning model will learn about these invalid labels. The model's learning capability significantly improves when trained with valid data, which provides stronger label signals, as opposed to invalid data that weakens these signals due to reduced exposure to valid label points.

Ensuring that the data is accurate

Even though the data is valid, it may not be accurate. Data accuracy measures the percentage of data that matches real-world data or verifiable sources. Considering the preceding example of the property area, to measure data accuracy, we may have to look up a reliable published dataset and check the population of the area and the type of the area. Let's assume that the population matches the verifiable data source, but the area type source is unavailable. Using the rule of what defines a rural area and what defines an urban area, we can measure data accuracy.

Using this business rule, we will create a new label called true_property_area that takes rural as a value when the population is 20,000 or below; otherwise, takes urban as a value:

```
df['true_property_area'] = df.population.apply(lambda value: 'rural'
if value <= 20000 else 'urban')
```

Next, we print the rows of the dataset to see if there are any mismatches between `property_area` and `true_property_area`:

```
df[['true_property_area', 'property_area', 'population']]
   true_property_area property_area  population
0               urban         urban       99539
1               urban         urban       78708
2               rural         urban        6192
3               urban         urban       99047
4               urban    semi_urban       51057
5               urban    semi_urban       74349
6               urban    semi_urban       22440
7               urban    semi_urban       99448
8               urban    semi_urban       21609
9               urban         rural       50100
```

Then, we sum the rows where `property_area` values match with the true values and divide this by the total number of rows to calculate data accuracy:

```
sum(df.property_area == df.true_property_area) / df.shape[0]
0.3
```

Instead of creating a function to calculate the accuracy, we can leverage `accuracy_score` from scikit-learn:

```
accuracy_score(y_pred=df.property_area, y_true=df.true_property_area)
0.3
```

As we can see, both methods return the same score. If inaccurate data enters the system, the model may learn inaccurately about semi-urban and rural areas, and at the time of inference, produce undesirable outcomes.

Ensuring that the data is fresh

Data freshness is another important aspect of measuring data quality that has an impact on the quality and robustness of machine learning applications. Let's imagine that we have a machine learning application that's been trained on 2019 and 2020 customer behavior and utilized to predict hotel room bookings up to April 2021. Maybe January and February numbers were quite accurate, but when March and April hit, accuracy dropped. This might have been due to COVID-19, something that was unseen by the data, and its effects were not captured. In machine learning, this is called data drift. This is happening here; the data distribution in March and April was quite different from the data distribution in 2019 and 2020. By ensuring that the data is fresh and up to date, we can train the model more regularly or as soon as data drift is detected.

To measure data drift, we will use the `alibi` Python package. However, there are more extensive Python packages that can help with this job. We recommend Evidently AI (https://www.evidentlyai.com/), a data and machine learning model monitoring toolkit, or WhyLogs (https://whylabs.ai/whylogs), an open source initiative by WhyLabs, to monitor model degradation and data drift.

Let's imagine that, on average, model accuracy starts degrading when the model is trained on data that is more than 5 days old and the model does poorly and starts costing the business when data is more than 10 days old. We want to be able to alert and capture when this happens. To demonstrate this scenario, we will create a sample dataset with a date column and define error and warning thresholds – that is, we print a warning if the data is 5 days old and block the application if the data is more than 10 days old. In practice, it is recommended to train the model on the most recent data available. Following a data-centric approach, we must encourage practitioners to have thresholds and **service-level agreements** (**SLAs**) defined with the providers of data so that they have mechanisms in place to request up-to-date data, penalize when SLAs are breached, and maybe reward when SLAs are met (encourage the importance of maintaining good-quality data).

Now we will generate 100 sample data points and demonstrate how to identify if data is stale using a date variable.

We utilize the `alibi` package to detect drift in the `loan_prediction` dataset. We will demonstrate the pitfalls of not detecting and acting on data drift by comparing accuracy before and after drift.

First, we import the `datetime` and `warning` packages:

```
from datetime import datetime, timedelta
import warnings
```

Next, we generate a base date – let's say the date when we run the code – and from the base date, generate 100 more dates in the past by subtracting one day at a time:

```
numdays = 100
base = datetime.today()
date_list = [base - timedelta(days=day) for day in range(numdays)] #
Subracting values from 1 to 100 from todays date
```

Then, we print the first 10 dates in the order these dates were generated, with the most recent being the first date:

```
[date.date().strftime('%Y-%m-%d') for date in date_list[0:10]]
['2023-02-04',
 '2023-02-03',
 '2023-02-02',
 '2023-02-01',
 '2023-01-31',
 '2023-01-30',
 '2023-01-29',
```

```
    '2023-01-28',
    '2023-01-27',
    '2023-01-26']
```

Next, we create a DataFrame with 100 rows by creating four columns. It will contain an `id` column with values from 1 to 100, a `date_loaded` column that contains the 100 dates we created previously, a `population` column with 100 random values between 1,000 and 100,000, and a `property_area` column with 40 values set to `urban`, 50 values set to `semi_urban,` and 10 values set to `rural`:

```python
np.random.seed(1)
data = {
    "id": np.linspace(start=1, stop=100, num=100, dtype=int),
    "population" : np.random.randint(low=1000, high=100000, size=100),
    "property_area": ["urban"]*40 + ["semi_urban"]*50 + ["rural"]*10,
    "date_loaded": date_list
}

df = pd.DataFrame(data=data)
```

Now, we visualize the first five data points:

```
df.head()
    id  population property_area                  date_loaded
0    1       99539         urban 2023-02-04 11:18:46.771142
1    2       78708         urban 2023-02-03 11:18:46.771142
2    3        6192         urban 2023-02-02 11:18:46.771142
3    4       99047         urban 2023-02-01 11:18:46.771142
4    5       51057         urban 2023-01-31 11:18:46.771142
```

Next, we create a one-line piece of code to demonstrate a way to subtract any date from today's date and extract the number of days between the dates:

```
(datetime.now() - df.date_loaded.max()).days
0
```

Then, we create a function that will accept a DataFrame and the date column of the DataFrame, and by default will issue a warning if data is more than 5 days old and block the application when data is more than 10 days old:

```python
def check_data_recency_days(df: pd.DataFrame, loaded_at_column: str,
   warning_at: int=5, error_at: int=10):
    """Function to detect data freshness"""

    df = df.copy()
```

```
    days_since_data_refreshed = (datetime.now() - df[loaded_at_
column].max()).days

    if days_since_data_refreshed < warning_at:
        print(f"Data is fresh and is {days_since_data_refreshed} days
old")

    elif error_at > days_since_data_refreshed >= warning_at:
        warnings.warn(f"Warning: Data is not fresh, and is {days_
since_data_refreshed} days old")

    else:
        raise ValueError(f"Date provided is too old and stale, please
contact source provider: {days_since_data_refreshed} days old")
```

Next, we run the function with the sample DataFrame we created previously. The function will state that the data is fresh and just 0 days old:

```
check_data_recency_days(df, "date_loaded")
Data is fresh and is 0 days old
```

To demonstrate the function's ability to issue a warning or error out when data is stale, we subset the data by removing 6 recent days and 12 recent days. We create two DataFrames – one that has 6 recent days removed and another that has 12 recent days removed. Then, we run the check_data_ recency_day function over these DataFrames. We see that when we run the function with 6-day-old data, the function will issue the warning, but when we run the function with 12-day-old data, the function will issue a Value error.

Let's create the two DataFrames:

```
df_filter_6_days = df[df.date_loaded <= (datetime.today()
- timedelta(days=6))]
df_filter_12_days = df[df.date_loaded <= (datetime.today()
- timedelta(days=12))]
```

Next, we run the function against the data that is 6 days old:

```
check_data_recency_days(df_filter_6_days, "date_loaded")
/var/folders/6f/p7312_7n4nq5hp35rfymms1h0000gn/T/
ipykernel_5374/1750573000.py:11: UserWarning: Warning: Data is not
fresh, and is 6 days old
  warnings.warn(f"Warning: Data is not fresh, and is {days_since_data_
refreshed} days old")
```

You can also run the function against the data that is 12 days old; it will generate similar output.

With that, we have demonstrated how to measure data staleness, catch warnings, and block the application when data is extremely stale. Next, we will demonstrate the impact of data staleness on a real-life dataset.

In real life, we wouldn't expect a company not to change its products or for consumer behavior not to change as newer products become available on the market. Companies have to constantly study changes in consumer behavior; otherwise, their performance degrades. Machine learning systems face the same issues as market forces change, data changes, and data distributions change. This has an impact on a model's performance if new data is quite different from the data it was trained on.

This is referred to as drift in machine learning and if it goes undetected and untreated, it can cause models to degrade.

Let's explore how to detect drift.

First, we import `TabularDrift` from the `alibi-detect` package:

```
import alibi
from alibi_detect.cd import TabularDrift
```

Next, we show the `TabularDrift` reference data, which is the data the machine learning system was trained on – in our case, the loan prediction transformed data before we passed it to the decision tree classifier. We also pass a value of `0.05` for the p-value test. If this value is breached by the test data distribution, the package will inform us that the test data has drifted from the training data:

```
cd = TabularDrift(x_ref=X_train_transformed.to_numpy(), p_val=.05 )
```

Now, we run the `predict` method to check if the test data has drifted. The `alibi` package utilizes the `Kolmogorov-Smirnov` test to determine if the two distributions differ. If the p-value exceeds 0.05, then the null hypothesis is rejected and it can be inferred that the `test` data distribution differs from the `train` data distribution. The output for this step will be `No`:

```
preds = cd.predict(X_test_transformed.to_numpy())
labels = ['No', 'Yes']
print('Drift: {}'.format(labels[preds['data']['is_drift']]))
Drift: No
```

Now, let's imagine that house prices started booming while incomes did not increase at the same rate. To simulate this scenario, we will increase the loan amount requested by 1.5 times the original test set, but increase the total income by 1.2 times the test set. Then, we update the new feature values that relied on the `loan_amount` and `income` variables:

```
 X_test_transformed['loan_amount'] = X_test_transformed['loan_
amount']*1.5
X_test_transformed['sum_applicant_income_coapplicant_income'] = X_
test_transformed['sum_applicant_income_coapplicant_income']*1.2
X_test_transformed.sum_applicant_income_coapplicant_income_div_loan_
amount = X_test_transformed.sum_applicant_income_coapplicant_income/X_
test_transformed.loan_amount
X_test_transformed.loan_amount_div_loan_amount_term = X_test_
transformed.loan_amount/X_test_transformed.loan_amount_term
```

Next, we rerun TabularDrift's `predict` method to check whether drift was detected. The output of this step is `Yes`:

```
preds = cd.predict(X_test_transformed.to_numpy())
labels = ['No', 'Yes']
print('Drift: {}'.format(labels[preds['data']['is_drift']]))
Drift: Yes
```

Then, we rerun the prediction on this drift-induced test data and check whether accuracy and ROC are affected:

```
testing_predictions_prob = model.predict_proba(X_test_transformed)
testing_predictions = model.predict(X_test_transformed)

testing_roc_auc = roc_auc_score(y_test, testing_predictions_prob[:,1])
testing_acc = accuracy_score(y_test, testing_predictions)

print(f"Testing roc is {testing_roc_auc} and testing_acc as {testing_
acc}")
Testing roc is 0.747858017135863 and testing_acc as 0.6935483870967742
```

As we can see, the distribution that was seen by the model when it was trained is different from the real data, and the impact is that model performance has degraded significantly. The ROC has dropped from 0.82 to 0.74 and accuracy has dropped from 82% to 70%. Hence, it's important to ensure that data is fresh and that as soon as drift is detected, the model is retrained with new data to ensure model performance does not deteriorate.

Summary

In this chapter, we gained a good understanding of the six key dimensions of data quality and why it's important to improve data quality for superior model performance. We further dived into the data-centric approach of improving model performance by iterating over the data, rather than iterating over various algorithms (model-centric approach), by improving the overall health of the data.

Next, we learned how to ensure data is consistent, unique, accurate, valid, fresh, and complete. We dived into various techniques of imputing missing values and when to apply which approach. We concluded that imputing missing values with machine learning can be better than using simple imputation methods, especially when data is MAR or MNAR. We also showed how to conduct error analysis and how to use the results to further improve model performance by either performing feature engineering, which involves building new features, or increasing the data size by creating synthetic data, which we will cover in the next chapter.

We also discussed why is it important to ensure data is fresh and not drifted from the original training set, and concluded that drifted data can hamper model performance.

Now that we have understood the importance of ensuring good-quality data across the six key dimensions of data quality, in the next chapter, we will dive into using synthetic data to further improve model performance, especially over edge cases. We will also dive into data augmentation, a technique that's used to create synthetic data for images so that algorithms can learn from more and better, especially when these new examples can come in various forms.

Techniques for Programmatic Labeling in Machine Learning

In machine learning, the accurate labeling of data is crucial for training effective models. Data labeling involves assigning meaningful categories or classes to data instances, and while traditionally a human-driven process, there are various programmatic approaches to dataset labeling. This chapter delves into the following methods of programmatic data labeling in machine learning:

- Pattern matching
- **Database (DB)** lookup
- Boolean flags
- Weak supervision
- Semi-weak supervision
- Slicing functions
- Active learning
- Transfer learning
- Semi-supervised learning

Technical requirements

To execute the code examples provided in this chapter on programmatic labeling techniques, ensure that you have the following technical prerequisites installed in your Python environment:

Python version

The examples in this chapter require Python version 3.7 or higher. You can check your Python version by running the following:

```
import sys
print(sys.version)
```

We recommend using the Jupyter Notebook **integrated development environment** (IDE) for an interactive and organized coding experience. If you don't have it installed, you can install it using this line:

```
pip install jupyter
```

Launch Jupyter Notebook with the following command:

```
jupyter notebook
```

Library requirements

Ensure that the following Python packages are installed in your environment. You can install them using the following commands:

```
pip install snorkel
pip install scikit-learn
pip install Pillow
pip install tensorflow
pip install pandas
pip install numpy
```

Additionally, for the TensorFlow and Keras components, you may need GPU support for optimal performance. Refer to the TensorFlow documentation for GPU installation instructions if you have a compatible GPU.

Pattern matching

In machine learning, one of the most important tasks is to label or classify data based on some criteria or patterns. However, labeling data manually can be time consuming and costly, especially when dealing with a large amount of data. By leveraging predefined patterns, this labeling approach enables the automatic assignment of meaningful categories or classes to data instances.

Pattern matching involves the identification of specific patterns or sequences within data that can be used as indicators for assigning labels. These patterns can be defined using regular expressions, rule-based systems, or other pattern recognition algorithms. The objective is to capture relevant information and characteristics from the data that can be matched against predefined patterns to infer labels accurately.

Pattern matching can be applied to various domains and scenarios in machine learning. Some common applications include the following:

- **Text classification**: In natural language processing, pattern matching can be utilized to label text data based on specific keywords, phrases, or syntactic patterns. This enables tasks such as sentiment analysis, spam detection, and topic categorization.

- **Image recognition**: Pattern matching can aid in labeling images by identifying distinctive visual patterns or features that correspond to specific classes. This technique can be valuable in tasks such as object recognition, facial detection, and image segmentation.

- **Time series analysis**: When dealing with time-dependent data, pattern matching can be employed to label sequences of events or patterns that occur over time. This is particularly useful in financial analysis, anomaly detection, and predicting stock market trends.

- **Fraud detection**: Pattern matching can play a crucial role in identifying fraudulent activities by matching suspicious patterns or anomalies against known fraud patterns. This technique can help in credit card fraud detection, network intrusion detection, and cybersecurity.

Pattern matching offers several advantages as a labeling technique in machine learning:

- **Automation and efficiency**: By automating the labeling process, pattern matching reduces the reliance on manual labeling, saving time and effort. It allows for large-scale dataset labeling with increased efficiency.

- **Flexibility and adaptability**: Patterns can be easily modified or extended to accommodate new data or evolving requirements. This provides flexibility in adapting to changing labeling criteria and ensures scalability.

- **Interpretability**: Pattern matching provides a transparent and interpretable approach to labeling, as the rules and patterns can be examined and understood. This aids in the transparency and explainability of the labeling process.

- **Complementing other techniques**: Pattern matching can be used in conjunction with other labeling techniques, such as weak supervision or transfer learning, to enhance the overall labeling accuracy and robustness of machine learning models.

While pattern matching is a valuable labeling technique, it also presents certain challenges and considerations:

- **Noise and ambiguity**: Data instances that do not perfectly match predefined patterns may introduce noise or ambiguity in the labeling process. Handling such cases requires careful design and consideration of pattern definitions.

- **Scalability**: As datasets grow larger, the scalability of pattern matching becomes crucial. Efficient algorithms and techniques must be employed to handle the increasing computational demands.

- **Overfitting**: Overfitting can occur if patterns are too specific and fail to generalize well to unseen data instances. Regularization techniques and cross-validation can be used to mitigate this risk.

In this section of the chapter, we will explore how to create pattern-matching labeling functions using Python and apply them to the `credit-g` dataset. The `credit-g` dataset, also known as the German Credit dataset, is a collection of data points used for risk analysis in the field of finance and machine learning. It's used to classify people as good or bad credit risks based on a set of attributes.

The dataset consists of 20 variables, including both numerical and categorical data. These variables provide information about each individual, such as their checking account status, credit history, purpose of the loan, credit amount, savings account/bonds, employment, installment rate in percentage of disposable income, personal status and gender, and other attributes.

Each entry in the dataset represents an individual who has applied for a loan. The target variable indicates whether the individual is classified as a 'good' or 'bad' credit risk. This makes the dataset particularly useful for supervised machine learning tasks, especially binary classification problems.

The `credit-g` dataset is widely used in academia and industry for developing and testing machine learning models for credit risk assessment. It is available on several platforms, such as DataHub, Kaggle, OpenML, and UCI Machine Learning Repository.

> **Note**
> Please note that the specifics of the variables might differ slightly depending on the source of the dataset.

We can start by loading the `credit-g` dataset into Python. The dataset contains information about loan applicants, including their demographic information, financial information, and loan approval status. We can use the `pandas` library to load the dataset and explore its structure:

```
from sklearn.datasets import fetch_openml
import pandas as pd

# Fetch the credit-g dataset
credit_g = fetch_openml(name='credit-g')

# Convert to DataFrame
df = pd.DataFrame(credit_g.data, columns=credit_g.feature_names)
target = pd.Series(credit_g.target)

# If you want to add the target variable into your DataFrame
df['target'] = target
```

```
# Show top rows of the credit-g dataset
df.head().T
```

Here are the first five rows of the dataset:

	0	1	2	3	4
checking_status	<0	0<=X<200	no checking	<0	<0
duration	6.0	48.0	12.0	42.0	24.0
credit_history	critical/other existing credit	existing paid	critical/other existing credit	existing paid	delayed previously
purpose	radio/tv	radio/tv	education	furniture/equipment	new car
credit_amount	1169.0	5951.0	2096.0	7882.0	4870.0
savings_status	no known savings	<100	<100	<100	<100
employment	>=7	1<=X<4	4<=X<7	4<=X<7	1<=X<4
installment_commitment	4.0	2.0	2.0	2.0	3.0
personal_status	male single	female div/dep/mar	male single	male single	male single
other_parties	none	none	none	guarantor	none
residence_since	4.0	2.0	3.0	4.0	4.0
property_magnitude	real estate	real estate	real estate	life insurance	no known property
age	67.0	22.0	49.0	45.0	53.0
other_payment_plans	none	none	none	none	none
housing	own	own	own	for free	for free
existing_credits	2.0	1.0	1.0	1.0	2.0
job	skilled	skilled	unskilled resident	skilled	skilled
num_dependents	1.0	1.0	2.0	2.0	2.0
own_telephone	yes	none	none	none	none
foreign_worker	yes	yes	yes	yes	yes
target	good	bad	good	good	bad

Figure 6.1 – The features (first column) and first five rows of the credit-g dataset

Now that we have loaded the dataset, we can create pattern-matching labeling functions. In this example, we will create two labeling functions that assign labels to loan applicants based on their income and credit history. income_labeling_function assigns a label of 1 to loan applicants with an income greater than 5,000 and a label of 0 to all others. credit_history_labeling_function assigns a label of 1 to loan applicants with a credit history of 1, and a label of 0 to all others.

Given the features in the credit-g dataset, we can create two labeling functions based on credit_amount and age. credit_amount_labeling_function assigns a label of 1 to loan applicants with a credit amount greater than 5,000 and a label of 0 to all others. age_labeling_function assigns a label of 1 to loan applicants older than 30 and a label of 0 to all others:

```
def credit_amount_labeling_function(df):
    if df["credit_amount"] > 5000:
        return 1
    else:
```

```
        return 0

def age_labeling_function(df):
    if df["age"] > 30:
        return 1
    else:
        return 0
```

After creating the labeling functions, we can apply them to the `credit-g` dataset. We can use the `apply` function in pandas to apply the labeling functions to each row of the dataset. The `apply` function applies the labeling functions to each row of the dataset and assigns the labels to new columns in the dataset:

```
df["credit_amount_label"] = df.apply(credit_amount_labeling_function,
axis=1)
df["age_label"] = df.apply(age_labeling_function, axis=1)
df.head().T
```

Here is the output DataFrame using these functions. The DataFrame now has two additional columns with newly created labels:

	0	1	2	3	4
checking_status	<0	0<=X<200	no checking	<0	<0
duration	6.0	48.0	12.0	42.0	24.0
credit_history	critical/other existing credit	existing paid	critical/other existing credit	existing paid	delayed previously
purpose	radio/tv	radio/tv	education	furniture/equipment	new car
credit_amount	1169.0	5951.0	2096.0	7882.0	4870.0
savings_status	no known savings	<100	<100	<100	<100
employment	>=7	1<=X<4	4<=X<7	4<=X<7	1<=X<4
installment_commitment	4.0	2.0	2.0	2.0	3.0
personal_status	male single	female div/dep/mar	male single	male single	male single
other_parties	none	none	none	guarantor	none
residence_since	4.0	2.0	3.0	4.0	4.0
property_magnitude	real estate	real estate	real estate	life insurance	no known property
age	67.0	22.0	49.0	45.0	53.0
other_payment_plans	none	none	none	none	none
housing	own	own	own	for free	for free
existing_credits	2.0	1.0	1.0	1.0	2.0
job	skilled	skilled	unskilled resident	skilled	skilled
num_dependents	1.0	1.0	2.0	2.0	2.0
own_telephone	yes	none	none	none	none
foreign_worker	yes	yes	yes	yes	yes
target	good	bad	good	good	bad
credit_amount_label	0	1	0	1	0
age_label	1	0	1	1	1

Figure 6.2 – The updated credit-g dataset with two new features, credit_amount_label and age_label

Having explored pattern-matching functions, we now shift our focus to the simplicity and effectiveness of database lookup techniques. In this next section, we'll harness structured databases to enhance labeling accuracy, making our approach even more robust.

Database lookup

The **database lookup** (**DB lookup**) labeling technique provides a powerful means of assigning labels to data instances by leveraging information stored in databases. By querying relevant databases and retrieving labeled information, this approach enables automated and accurate labeling. This technique involves searching and retrieving labels from databases based on specific attributes or key-value pairs associated with data instances. It relies on the premise that databases contain valuable labeled information that can be utilized for data labeling purposes. By performing queries against databases, relevant labels are fetched and assigned to the corresponding data instances.

The DB lookup technique finds application in various domains and scenarios within machine learning. Some common applications include the following:

- **Entity recognition**: In natural language processing tasks, such as named entity recognition or entity classification, DB lookup can be used to retrieve labels for entities based on their attributes stored in databases. This aids in the accurate identification and categorization of entities in text data.

- **Product categorization**: E-commerce platforms often maintain databases containing product information, including categories and attributes. DB lookup can be employed to fetch product labels based on their features, allowing for automated categorization and organization of products.

- **Geospatial analysis**: Databases containing geographical information, such as maps or geotagged data, can be queried using DB lookup to assign labels based on spatial attributes. This technique facilitates tasks such as location-based recommendations, geospatial clustering, and boundary identification.

- **Medical diagnosis**: Medical databases store extensive information about diseases, symptoms, and patient records. DB lookup can be utilized to retrieve relevant labels for patient symptoms, aiding in automated medical diagnosis and decision support systems.

Now, let's talk about Boolean flag labeling. It's a simple yet powerful method that helps us improve and automate labeling by using clear and logical conditions.

Boolean flags

The **Boolean flags** labeling technique involves the use of binary indicators to assign labels to data instances. These indicators, often represented as Boolean variables (`true`/`false` or `1`/`0`), are associated with specific characteristics or properties that help identify the desired label. By examining the presence or absence of these flags, data instances can be automatically labeled.

The Boolean flags labeling technique finds applications across various domains in machine learning. Some common applications include the following:

- **Data filtering**: Boolean flags can be used to filter and label data instances based on specific criteria. For example, in sentiment analysis, a positive sentiment flag can be assigned to text instances that contain positive language or keywords, while a negative sentiment flag can be assigned to instances with negative language.

- **Event detection**: Boolean flags can aid in labeling instances to detect specific events or conditions. For instance, in cybersecurity, a flag can be set to indicate instances with suspicious network activity, enabling the identification of potential security threats.

- **Anomaly detection**: Boolean flags can be used to label instances as normal or anomalous. By defining flags that capture typical patterns or behaviors, instances that deviate from these patterns can be flagged as anomalies, facilitating anomaly detection tasks.

- **Quality control**: Boolean flags can assist in labeling instances for quality control purposes. For example, in manufacturing, flags can be set to label instances as defective or non-defective based on predefined quality criteria.

The Boolean flags labeling technique offers several advantages in machine learning applications:

- **Simplicity and efficiency**: Boolean flags provide a straightforward and efficient labeling mechanism. The labeling process involves checking the presence or absence of flags, which can be implemented using simple conditional statements or logical operations.

- **Flexibility and customization**: Boolean flags allow for customization and adaptability to different labeling scenarios. Flags can be defined based on specific criteria or requirements, providing flexibility in assigning labels according to the desired characteristics.

- **Interpretability**: The Boolean flags labeling technique offers interpretability, as the presence or absence of flags directly corresponds to the assigned labels. This transparency allows for better understanding and validation of the labeling process.

- **Scalability**: Boolean flags can be easily scaled to handle large datasets. Since the labeling decision is based on binary indicators, the computational overhead remains low, making it suitable for processing massive amounts of data.

While the Boolean flags labeling technique provides simplicity and efficiency, certain challenges and considerations should be taken into account:

- **Feature engineering**: Designing effective Boolean flags requires careful feature engineering. The flags should be informative and relevant to the desired labels, necessitating a deep understanding of the problem domain and data characteristics.

- **Data imbalance**: In scenarios where the data is imbalanced, meaning one label dominates over others, the Boolean flags technique may face challenges. Proper handling techniques, such as oversampling or under-sampling, may be required to address the imbalance issue.

- **Generalization**: Boolean flags may not capture the full complexity of the underlying data distribution, potentially leading to overfitting or limited generalization. It is important to consider complementary techniques, such as feature extraction or more advanced machine learning algorithms, to enhance the performance and generalization capabilities.

- **Flag interpretation**: While Boolean flags provide interpretability, it is crucial to carefully interpret the flags' meanings in relation to the assigned labels. In some cases, the flags may capture correlations rather than causal relationships, requiring further investigation for a more accurate interpretation.

You may have already noticed some similarities between Boolean flags and one-hot encoding (covered in *Chapter 5, Techniques for Data Cleaning*). Therefore, it's important to understand when these techniques are appropriate.

When choosing between Boolean flags and one-hot encoding, the specific use case is a crucial factor. If you're working with a categorical variable that can be naturally divided into two categories or states (such as yes/no, true/false), using a Boolean flag might be the best option. It's simpler, more memory-efficient, and can make the model easier to interpret.

For example, if you're predicting whether an email is spam or not, a Boolean flag such as `contains_link` (1 if the email contains a link, 0 otherwise) could be a very effective feature. This simplicity can lead to more interpretable models, as each feature directly corresponds to a condition or state.

On the other hand, one-hot encoding is more suitable for categorical variables with multiple categories where no natural binary division exists. For instance, if you're working with a feature such as `color` with values such as `red`, `blue`, `green`, etc., one-hot encoding would be a better choice. That's because the numbers assigned to each category shouldn't imply a mathematical relationship between the categories unless one exists. For example, encoding red as 1 and blue as 2 doesn't mean blue is twice red.

To avoid implying such unintended relationships, creating a separate feature for each possible color is preferred. This approach captures more information about the color feature and doesn't impose an arbitrary order or importance on the different colors.

Furthermore, the type of machine learning model being used also influences the choice. Some models, such as decision trees and random forests, can handle categorical variables quite well, so one-hot encoding (which increases the dimensionality of the dataset) might not be necessary. However, others, such as linear regression, logistic regression, and support vector machines, require numerical input, necessitating some form of encoding for categorical variables.

Lastly, it's worth noting that these aren't the only methods for handling categorical data. There are other techniques, such as ordinal encoding, target encoding, and bin counting, each with its own strengths and weaknesses. The key is to understand the nature of your data and the requirements of your specific use case to choose the most appropriate method.

Let's explore how to utilize Boolean flags in Python with the `credit-g` dataset. Imagine we want to create a function that applies Boolean flags to label data points according to basic rules or heuristics. For instance, we can write a function that evaluates whether a credit applicant's credit amount is above a specific threshold, subsequently assigning a Boolean flag to the data point based on this assessment.

The following functions will check if the credit amount is below or above the median credit amount:

```
def lf_credit_amount_above_median(df):
    credit_amount_median = df['credit_amount'].median()
    return df['credit_amount'] >= credit_amount_median
```

Now that we have defined our function, we can apply it to our `df` DataFrame to label the data points with Boolean flags:

```
df['LF_CreditAmountAboveMedian'] = lf_credit_amount_above_median(df)
df.head().T
```

Figure 6.3 is the output DataFrame after we have applied these functions. Notice that we have now created a new column giving us additional information on the applicant's credit amount, which can be used as a feature in machine learning models.

	0	1	2	3	4
checking_status	<0	0<=X<200	no checking	<0	<0
duration	6.0	48.0	12.0	42.0	24.0
credit_history	critical/other existing credit	existing paid	critical/other existing credit	existing paid	delayed previously
purpose	radio/tv	radio/tv	education	furniture/equipment	new car
credit_amount	1169.0	5951.0	2096.0	7882.0	4870.0
savings_status	no known savings	<100	<100	<100	<100
employment	>=7	1<=X<4	4<=X<7	4<=X<7	1<=X<4
installment_commitment	4.0	2.0	2.0	2.0	3.0
personal_status	male single	female div/dep/mar	male single	male single	male single
other_parties	none	none	none	guarantor	none
residence_since	4.0	2.0	3.0	4.0	4.0
property_magnitude	real estate	real estate	real estate	life insurance	no known property
age	67.0	22.0	49.0	45.0	53.0
other_payment_plans	none	none	none	none	none
housing	own	own	own	for free	for free
existing_credits	2.0	1.0	1.0	1.0	2.0
job	skilled	skilled	unskilled resident	skilled	skilled
num_dependents	1.0	1.0	2.0	2.0	2.0
own_telephone	yes	none	none	none	none
foreign_worker	yes	yes	yes	yes	yes
target	good	bad	good	good	bad
credit_amount_label	0	1	0	1	0
age_label	1	0	1	1	1
LF_CreditAmountAboveMedian	False	True	False	True	True

Figure 6.3 – The credit-g dataset with the new Boolean flag LF_CreditAmountAboveMedian added

In the next section, let's explore weak supervision—a sophisticated labeling technique that adeptly integrates information from various sources, navigating the intricacies of real-world data to enhance precision and adaptability.

Weak supervision

Weak supervision is a labeling technique in machine learning that leverages imperfect or noisy sources of supervision to assign labels to data instances. Unlike traditional labeling methods that rely on manually annotated data, weak supervision allows for a more scalable and automated approach to labeling. It refers to the use of heuristics, rules, or probabilistic methods to generate approximate labels for data instances.

Rather than relying on a single authoritative source of supervision, weak supervision harnesses multiple sources that may introduce noise or inconsistency. The objective is to generate labels that are "weakly" indicative of the true underlying labels, enabling model training in scenarios where obtaining fully labeled data is challenging or expensive.

For instance, consider a task where we want to build a machine learning model to identify whether an email is spam or not. Ideally, we would have a large dataset of emails that are accurately labeled as "spam" or "not spam." However, obtaining such a dataset could be challenging, time-consuming, and expensive.

With weak supervision, we can use alternative, less perfect ways to label our data. For instance, we could create some rules or heuristics based on common patterns in spam emails. Here are a few examples of such rules:

- If the email contains words such as "lottery", "win", or "prize", it might be spam
- If the email is from an unknown sender and contains many links, it might be spam
- If the email contains phrases such as "urgent action required", it might be spam

Using these rules, we can automatically label our email dataset. These labels won't be perfect— there will be false positives (non-spam emails incorrectly labeled as spam) and false negatives (spam emails incorrectly labeled as non-spam). But they give us a starting point for training our machine learning model.

The model can then learn from these "weak" labels and, with a sufficiently large and diverse dataset, should still be able to generalize well to new, unseen emails. This makes weak supervision a scalable and efficient approach to labeling, particularly useful when perfect labels are hard to come by.

Weak supervision can be derived from various sources, including the following:

- **Rule-based systems**: Domain experts or heuristics-based approaches can define rules or guidelines for labeling data based on specific patterns, features, or conditions. These rules may be derived from knowledge bases, existing models, or expert opinions.
- **Crowdsourcing**: Leveraging the power of human annotators through crowdsourcing platforms, weak supervision can be obtained by aggregating the annotations from multiple individuals. This approach introduces noise but can be cost-effective and scalable.
- **Distant supervision**: Distant supervision involves using existing labeled data that may not perfectly align with the target task but can serve as a proxy. An example is using existing data with auxiliary labels to train a model for a related but different task.
- **Data augmentation**: Weak supervision can be obtained through data augmentation techniques such as data synthesis, transformation, or perturbation. By generating new labeled instances based on existing labeled data, weak supervision can be expanded.

Weak supervision offers several advantages in machine learning applications:

- **Scalability**: Weak supervision allows for large-scale data labeling by leveraging automated or semi-automated techniques. It reduces the manual effort required for manual annotation, enabling the utilization of larger datasets.

- **Cost-effectiveness**: By leveraging weakly supervised sources, the cost of obtaining labeled data can be significantly reduced compared to fully supervised approaches. This is particularly beneficial in scenarios where manual labeling is expensive or impractical.

- **Flexibility and adaptability**: Weak supervision techniques can be easily adapted and modified to incorporate new sources of supervision or update existing rules. This flexibility allows for iterative improvement and refinement of the labeling process.

- **Handling noisy labels**: Weak supervision techniques can handle noisy or inconsistent labels by aggregating multiple weak signals. This robustness to noise reduces the impact of individual labeling errors on the overall training process.

There are, however, certain challenges and considerations to be aware of:

- **Noise and label quality**: Weakly supervised labels may contain noise or errors due to the imperfect nature of the supervision sources. Careful evaluation and validation are necessary to ensure label quality and minimize the propagation of noisy labels during model training.

- **Trade-off between precision and recall**: Weak supervision techniques often prioritize scalability and coverage over precision. Balancing the trade-off between recall (coverage) and precision (accuracy) is essential in obtaining reliable weakly labeled data.

- **Labeling confidence and model training**: Handling the uncertainty associated with weakly supervised labels is crucial. Techniques such as label calibration, data augmentation, or active learning can be employed to mitigate the impact of label uncertainty during model training.

- **Generalization and model performance**: Weakly supervised models may struggle with generalizing to unseen or challenging instances due to the inherent noise in the labels. Strategies such as regularization, ensemble methods, or transfer learning can be employed to enhance model performance.

In this section, we will explore how to use labeling functions in Python to train a machine learning model on the *Loan Prediction* dataset we introduced in *Chapter 5, Techniques for Data Cleaning*. First, we need to prepare the data by importing the necessary libraries and loading the dataset, and we need to preprocess the data by handling missing values and encoding categorical variables:

```
import pandas as pd
import numpy as np

df = pd.read_csv('train_loan_prediction.csv')
df['LoanAmount'].fillna(df['LoanAmount'].mean(), inplace=True)
```

```
df['Credit_History'].fillna(df['Credit_History'].mode()[0],
inplace=True)
df['Self_Employed'].fillna('No',inplace=True)
df['Gender'].fillna(df['Gender'].mode()[0], inplace=True)
df['Married'].fillna(df['Married'].mode()[0], inplace=True)
df['Dependents'].fillna(df['Dependents'].mode()[0], inplace=True)
df['Loan_Amount_Term'].fillna(df['Loan_Amount_Term'].mode()[0],
inplace=True)
df['Credit_History'].fillna(df['Credit_History'].mode()[0],
inplace=True)
```

We will use the `LabelEncoder` function from scikit-learn's `preprocessing` class to encode categorical columns:

```
from sklearn.preprocessing import LabelEncoder

cat_features = ['Gender', 'Married','Dependents', 'Education', 'Self_
Employed', 'Property_Area']
for feature in cat_features:
    encoder = LabelEncoder()
    df[feature] = encoder.fit_transform(df[feature])
```

Now, we can define our labeling functions. In this example, we will define three labeling functions based on some simple heuristics. These labeling functions take in a row of the dataset as input and return a label. The label is 1 if the row is likely to be in the positive class, 0 if it is likely to be in the negative class, and -1 if it is uncertain. Functions such as these are commonly used in weak supervision approaches where you have a large amount of unlabeled data and you want to generate noisy labels for them:

```
from snorkel.labeling import labeling_function
@labeling_function()
def lf1(df):
    if df['Education'] == 0:
        return 0
    elif df['Self_Employed'] == 0:
        return 1
    else:
        return -1
@labeling_function()
def lf2(df):
    if df['Credit_History'] == 1:
        if df['LoanAmount'] <= 120:
            return 1
        else:
            return 0
    else:
```

```
            return -1
@labeling_function()
def lf3(df):
    if df['Married'] == 1:
        if df['Dependents'] == 0:
            return 1
        elif df['Dependents'] == 1:
            return 0
        else:
            return -1
    else:
        return -1
```

We can apply the labeling functions to the dataset using the Snorkel library. Here, we create a list of the three labeling functions and use the `PandasLFApplier` to apply them to the dataset. The output is a `L_train` matrix where each row corresponds to a data point and each column corresponds to a labeling function:

```
LFs = [lf1, lf2, lf3]

from snorkel.labeling import PandasLFApplier

applier = PandasLFApplier(lfs=LFs)
L_train = applier.apply(df)
```

You'll see the following output:

```
100%|███████████████████████████████████████████████| 614/614 [00:00<00:00, 11616.06it/s]
```

Figure 6.4 – Progress bar showing the training progress

To improve the output of the labeling functions, we need to combine them to obtain a more accurate label for each data point. We can do this using the `LabelModel` class in the Snorkel library. The `LabelModel` class is a probabilistic model used for combining the outputs of multiple labeling functions to generate more accurate and reliable labeling for each data point. It plays a crucial role in addressing the noise and inaccuracies that may arise from individual labeling functions. We create a `LabelModel` object as `label_model` and fit it to the output of the labeling functions. The cardinality parameter specifies the number of classes, which is 2 in this case. We also specify the number of epochs to train for and a random seed for reproducibility:

```
from snorkel.labeling.model import LabelModel
from snorkel.labeling import PandasLFApplier, LFAnalysis
from sklearn.metrics import accuracy_score
```

```
label_model = LabelModel(cardinality=2, verbose=True)
label_model.fit(L_train=L_train, n_epochs=500, log_freq=100, seed=123)
```

After executing the preceding code snippet utilizing Snorkel's labeling model, a progress bar will display the incremental application of the labeling:

```
INFO:root:Computing O...
INFO:root:Estimating \mu...
  0%|                                                               | 0/500 [00:00<?, ?epoch/s]INFO:
root:[0 epochs]: TRAIN:[loss=0.791]
 13%|█████████                                                      | 63/500 [00:00<00:00, 619.30epoch/s]INF
O:root:[100 epochs]: TRAIN:[loss=0.001]
 38%|███████████████████████████                                   | 191/500 [00:00<00:00, 632.46epoch/s]INFO:
root:[200 epochs]: TRAIN:[loss=0.000]
 51%|████████████████████████████████████                          | 256/500 [00:00<00:00, 637.50epoch/s]INF
O:root:[300 epochs]: TRAIN:[loss=0.000]
 78%|███████████████████████████████████████████████████████       | 390/500 [00:00<00:00, 649.72epoch/s]INF
O:root:[400 epochs]: TRAIN:[loss=0.000]
100%|██████████████████████████████████████████████████████████████| 500/500 [00:00<00:00, 638.65epoch/s]
INFO:root:Finished Training
```

Figure 6.5 – The Snorkel progress bar showing the incremental progress of the labeling process

We can now use the `LabelModel` class to generate labels for the training data and evaluate its performance:

```
probs_train = label_model.predict_proba(L_train)
mapping = {'Y': 1, 'N': 0}
Y_train = df['Loan_Status'].map(mapping).values
score_train = label_model.score(L_train, Y_train)
df['label'] = probs_train.argmax(axis=1)
df['label'] = df['label'].map({1: 'Y', 0: 'N'})
print(f"Accuracy: {accuracy_score(df['Loan_Status'].values,
df['label'].values)}")
```

Now, let's learn how to use semi-weak supervision for labeling. It's a smart technique that combines weak supervision with a bit of manual labeling to make our labels more accurate.

Semi-weak supervision

Semi-weak supervision is a technique used in machine learning to improve the accuracy of a model by combining a small set of labeled data with a larger set of weakly labeled data. In this approach, the labeled data is used to guide the learning process, while the weakly labeled data provides additional information to improve the accuracy of the model.

Semi-weak supervision is particularly useful when labeled data is limited or expensive to obtain and can be applied to a wide range of machine learning tasks, such as text classification, image recognition, and object detection.

In the loan prediction dataset, we have a set of data points representing loan applications, each with a set of features such as income, credit history, and loan amount, and a label indicating whether the loan was approved or not. However, this labeled data may be incomplete or inaccurate, which can lead to poor model performance.

To address this issue, we can use semi-weak supervision to generate additional labels for the loan prediction dataset. One approach is to use weak supervision techniques to generate labels automatically based on heuristics or rules. For example, we can use regular expressions to identify patterns in the loan application text data that are indicative of a high-risk loan. We can also use external data sources, such as credit reports or social media data, to generate additional weakly labeled data.

Once we have a set of weakly labeled data, we can use it to train a model along with the small set of labeled data. The labeled data is used to guide the learning process, while the weakly labeled data provides additional information to improve the accuracy of the model. By using semi-weak supervision, we can effectively use all available data to improve model performance.

Here is an example of how to implement semi-weak supervision for the loan prediction dataset using Snorkel and Python. We first import the necessary libraries and functions from those libraries. We then load the dataset using pandas:

```
import pandas as pd
import numpy as np
from sklearn.model_selection import train_test_split
from snorkel.labeling import labeling_function
from snorkel.labeling import PandasLFApplier
from snorkel.labeling import LFAnalysis
from snorkel.labeling.model import LabelModel
df = pd.read_csv('train_loan_prediction.csv')
```

Let's define a function to preprocess the dataset. We will use similar preprocessing methods to the ones discussed earlier in this chapter:

```
def preprocess_data(df):
    df['Gender'] = df['Gender'].fillna('Unknown')
    df['Married'] = df['Married'].fillna('Unknown')
    df['Dependents'] = df['Dependents'].fillna('0')
    df['Self_Employed'] = df['Self_Employed'].fillna('Unknown')
    df['LoanAmount'] = df['LoanAmount'].fillna(df['LoanAmount'].
mean())
    df['Loan_Amount_Term'] = df['Loan_Amount_Term'].fillna(df['Loan_
Amount_Term'].mean())
    df['Credit_History'] = df['Credit_History'].fillna(-1)
    df['LoanAmount_bin'] = pd.cut(df['LoanAmount'], bins=[0, 100, 200,
700], labels=['Low', 'Average', 'High'])
    df['TotalIncome'] = df['ApplicantIncome'] +
df['CoapplicantIncome']
```

```
    df['TotalIncome_bin'] = pd.cut(df['TotalIncome'], bins=[0, 2500,
4000, 6000, 81000], labels=['Low', 'Average', 'High', 'Very high'])
    df = df.drop(['Loan_ID', 'ApplicantIncome', 'CoapplicantIncome',
'LoanAmount', 'TotalIncome'], axis=1)
    return df
```

Now we create three labeling functions for ApplicantIncome and LoanAmount. The lf1(x) function takes a data instance x as input and performs a labeling operation based on the value of the ApplicantIncome feature. If the ApplicantIncome is less than 5,000, the function returns a label of 0. Otherwise, if the ApplicantIncome is greater than or equal to 5,000, the function returns a label of 1. Essentially, this function assigns a label of 0 to instances with low applicant income and a label of 1 to instances with higher applicant income.

The lf2(x) function also takes a data instance x as input and assigns a label based on the value of the LoanAmount feature. If the LoanAmount is greater than 200, the function returns a label of 0. Conversely, if the LoanAmount is less than or equal to 200, the function returns a label of 1. This function categorizes instances with large loan amounts as label 0 and instances with smaller loan amounts as label 1.

Utilizing the lf3(x) function, we compute the ratio between the loan amount and the applicant's income. This ratio serves as a crucial metric in determining the feasibility of the loan. Based on this calculated ratio, we categorize the data points into different labels. If the loan-to-income ratio falls below or equals 0.3, we assign a label of 1, indicating approval of the loan request. In cases where the ratio exceeds 0.3 but remains less than or equal to 0.5, we designate the data point with a label of 0, signifying uncertainty regarding loan approval. Conversely, if the ratio surpasses 0.5, we assign the label -1, indicating denial of the loan application. This approach enables us to incorporate the affordability aspect into our labeling process, enhancing the granularity of our weak supervision approach for loan approval prediction:

```
@labeling_function()
def lf1(x):
    return 0 if x['ApplicantIncome'] < 5000 else 1


@labeling_function()
def lf2(x):
    return 0 if x['LoanAmount'] > 200 else 1


@labeling_function()
def lf3(x):
    # Calculate the ratio of loan amount to applicant's income
    loan_to_income_ratio = x['LoanAmount'] / x['ApplicantIncome']
```

```
# Return label based on the loan-to-income ratio
if loan_to_income_ratio <= 0.3:
    return 1  # Approve loan
elif loan_to_income_ratio > 0.3 and loan_to_income_ratio <= 0.5:
    return 0  # Label as uncertain
else:
    return -1  # Deny loan
```

We then apply the preprocessing techniques to the input data (df). The preprocess_data function is used to perform the necessary preprocessing steps. The resulting preprocessed data is stored in the variable X. Additionally, the target variable, Loan_Status, is transformed from categorical values (N and Y) to numerical values (0 and 1) and stored in the variable y. This step ensures that the data is ready for training and evaluation:

```
X = preprocess_data(df.drop('Loan_Status', axis=1))
y = df['Loan_Status'].replace({'N': 0, 'Y': 1})
```

The next step involves splitting the preprocessed data into training and testing sets. The train_test_split function from the scikit-learn library is used for this purpose. The data is divided into X_train and X_test for the features and y_train and y_test for the corresponding labels. This separation allows for training the model on the training set and evaluating its performance on the test set:

```
X_train, X_test, y_train, y_test = train_test_split(X, y, test_
size=0.2, random_state=42)
```

Now we apply the two labeling functions, lf1 and lf2, to the training set (X_train) using the PandasLFApplier class. The resulting weakly labeled data is stored in L_train_weak. The LFs analyze the features of each instance and assign labels based on predefined rules or conditions:

```
lfs = [lf1, lf2, lf3]
applier = PandasLFApplier(lfs)
L_train_weak = applier.apply(X_train)
```

The label model is instantiated using the LabelModel class. It is configured with a cardinality of 2 (indicating binary classification) and set to run in verbose mode for progress updates. The label model is then trained on the training data (L_train_weak) using the fit method:

```
label_model = LabelModel(cardinality=2, verbose=True)
label_model.fit(L_train_weak)
```

Once the label model is trained, it is evaluated on the test set (`X_test`) to assess its performance. The applier object is used again to apply the labeling functions to the test set, resulting in `L_test`, which contains the weakly labeled instances. The score method of the label model is then used to calculate the accuracy of the label predictions compared to the ground truth labels (`y_test`):

```
L_test = applier.apply(X_test)
accuracy = label_model.score(L_test, y_test)["accuracy"]
print(f'Test accuracy: {accuracy:.3f}')
```

In the upcoming section, we explore slicing functions for labeling—an advanced technique that allows us to finely segment our data. These functions provide a tailored approach, enabling us to apply specific labeling strategies to distinct subsets of our dataset.

Slicing functions

Slicing functions are functions that operate on data instances and produce binary labels based on specific conditions. Unlike traditional labeling functions that provide labels for the entire dataset, slicing functions are designed to focus on specific subsets of the data. These subsets, or slices, can be defined based on various features, patterns, or characteristics of the data. Slicing functions offer a fine-grained approach to labeling, enabling more targeted and precise labeling of data instances.

Slicing functions play a crucial role in weak supervision approaches, where multiple labeling sources are leveraged to assign approximate labels. Slicing functions complement other labeling techniques, such as rule-based systems or crowdsourcing, by capturing specific patterns or subsets of the data that may be challenging to label accurately using other methods. By applying slicing functions to the data, practitioners can exploit domain knowledge or specific data characteristics to improve the labeling process.

To fully understand the concept of slicing functions, let's use an example of a dataset containing reviews for a range of products from an e-commerce website. Our goal is to label these reviews as either positive or negative.

For simplicity, let's consider two slicing functions:

- **Slicing Function 1 (SF1)**: This function targets reviews that contain the word "refund". It labels a review as negative if it includes the word "refund" and leaves it unlabeled otherwise. The intuition behind this slicing function is that customers asking for a refund are likely dissatisfied with their purchase, hence the negative sentiment.

- **Slicing Function 2 (SF2)**: This function targets reviews from customers who purchased electronics. It labels a review as positive if it includes words such as "great", "excellent", or "love" and labels it as negative if it includes words such as "broken", "defective", or "useless". It leaves the review unlabeled if it doesn't meet any of these conditions.

You will notice that these slicing functions operate on specific subsets of the data and enable us to incorporate domain knowledge into the labeling process. Therefore, designing effective slicing functions requires a combination of domain knowledge, feature engineering, and experimentation. Here are some key considerations for designing and implementing slicing functions:

- **Identify relevant slices**: Determine the specific subsets or slices of the data that are relevant to the labeling task. This involves understanding the problem domain, analyzing the data, and identifying distinct patterns or characteristics.

- **Define slicing conditions**: Specify the conditions or rules that capture the desired subsets of the data. These conditions can be based on feature thresholds, pattern matching, statistical properties, or any other relevant criteria.

- **Evaluate and iterate**: Assess the performance of the slicing functions by comparing the assigned labels to ground truth labels or existing labeling sources. Iterate on the design of the slicing functions, refining the conditions and rules to improve the quality of the assigned labels.

Slicing functions offer several benefits in the labeling process:

- **Fine-grained labeling**: Slicing functions allow for targeted labeling of specific subsets of the data, providing more detailed and granular labels that capture distinct patterns or characteristics.

- **Domain knowledge incorporation**: Slicing functions enable the incorporation of domain expertise and specific domain knowledge into the labeling process. This allows for more informed and context-aware labeling decisions.

- **Complementarity with other techniques**: Slicing functions complement other labeling techniques by capturing slices of the data that may be challenging to label using traditional methods. They provide an additional source of weak supervision that enhances the overall labeling process.

- **Scalability and efficiency**: Slicing functions can be automated and applied programmatically, allowing for scalable and efficient labeling of large datasets. This reduces the dependency on manual annotation and enables the labeling of data at a larger scale.

Let's understand how we can implement slicing functions in Python using the loan prediction dataset. We first import the required libraries and load the dataset into a pandas DataFrame. We will use the same preprocessing step discussed in the previous sections:

```
from snorkel.labeling import labeling_function
from snorkel.labeling import PandasLFApplier
from snorkel.labeling.model import LabelModel
from snorkel.labeling import LFAnalysis
df = pd.read_csv('train_loan_prediction.csv')
X = preprocess_data(df.drop('Loan_Status', axis=1))
y = df['Loan_Status'].replace({'N': 0, 'Y': 1})
X_train, X_test, y_train, y_test = train_test_split(X, y, test_
size=0.2, random_state=42)
```

To create slicing functions, we utilize the `@labeling_function` decorator provided by Snorkel. These functions encapsulate the labeling logic based on specific conditions or rules. For example, we can define slicing functions based on the `ApplicantIncome`, `LoanAmount`, or `Self_Employed` features:

```
@labeling_function()
def slice_high_income(x):
    return 1 if x['ApplicantIncome'] > 8000 else 0


@labeling_function()
def slice_low_income_high_loan(x):
    return 1 if x['ApplicantIncome'] < 4000 and x['LoanAmount'] > 150
else 0


@labeling_function()
def slice_self_employed(x):
    return 1 if x['Self_Employed'] == 'Yes' else 0
```

To apply the slicing functions to the training data, we use the `PandasLFApplier` class provided by Snorkel. This class takes the slicing functions as input and applies them to the training dataset, generating weak labels. The resulting weak labels will be used to train the label model later:

```
lfs = [slice_high_income, slice_low_income_high_loan, slice_self_
employed]
applier = PandasLFApplier(lfs)
L_train = applier.apply(df=X_train)
```

Once we have the weak labels from the slicing functions, we can train a label model using the `LabelModel` class from Snorkel. The label model learns the correlation between the weak labels and the true labels and estimates the posterior probabilities for each data instance. In this step, we create a `LabelModel` object, specify the cardinality of the labels (e.g., binary classification), and fit it to the weakly labeled training data:

```
label_model = LabelModel(cardinality=2, verbose=True)
label_model.fit(L_train, n_epochs=500, seed=42)
```

After training the label model, we want to evaluate its performance on the test data. We use the `PandasLFApplier` to apply the slicing functions to the test dataset, obtaining the weak labels. Then, we calculate the accuracy of the label model's predictions compared to the true labels of the test set:

```
L_test = applier.apply(df=X_test)
accuracy = label_model.score(L=L_test, Y=y_test)
print(f'Test accuracy: {accuracy["accuracy"]:.3f}')
```

Snorkel provides the `LFAnalysis` module, which allows us to analyze the performance and characteristics of the labeling functions. We can compute various metrics such as coverage, conflicts, and accuracy for each labeling function to gain insights into their effectiveness and potential issues:

```
LFAnalysis(L=L_train, lfs=lfs).lf_summary()
```

This will generate the following summary table:

	j	Polarity	Coverage	Overlaps	Conflicts
slice_high_income	0	[0, 1]	1.0	1.0	0.303462
slice_low_income_high_loan	1	[0, 1]	1.0	1.0	0.303462
slice_self_employed	2	[0, 1]	1.0	1.0	0.303462

Figure 6.6 – Summary table showing statistics for each labeling function (LF)

In the next section, we'll explore active learning for labeling—a clever strategy that involves picking the right data to label, making our model smarter with each iteration.

Active learning

In this section, we will explore the concept of active learning and its application in data labeling. Active learning is a powerful technique that allows us to label data more efficiently by actively selecting the most informative samples for annotation. By strategically choosing which samples to label, we can achieve higher accuracy with a smaller dataset, all else being equal. On the following pages, we will discuss various active learning strategies and implement them using Python code examples.

Active learning is a semi-supervised learning approach that involves iteratively selecting a subset of data points for manual annotation based on their informativeness. The key idea is to actively query the labels of the most uncertain or informative instances to improve the learning process. This iterative process of selecting and labeling samples can significantly reduce the amount of labeled data required to achieve the desired level of performance.

Let's start with a simple example of active learning to help you get the basic idea before going into detail on specific active learning strategies.

Suppose we are building a machine learning model to classify emails into spam and not spam. We have a large dataset of unlabeled emails, but manually labeling all of them would be very time-consuming.

Here is where active learning comes in:

1. **Initial training**: We start by randomly selecting a small subset of emails and manually labeling them as spam or not spam. We then train our model on this small labeled dataset.

2. **Uncertainty sampling**: After training, we use the model to make predictions on the rest of the unlabeled emails. However, instead of labeling all the emails, we choose the ones where the model is most uncertain about its predictions. For example, if our model outputs a probability close to 0.5 (i.e., it's unsure whether the email is spam or not), these emails are considered 'informative' or 'uncertain'.

3. **Label query**: We then manually label these uncertain emails, adding them to our training set.

4. **Iterative learning**: *Step 2* and *step 3* are repeated in several iterations— retraining the model with the newly labeled data, using it to predict labels for the remaining unlabeled data, and choosing the most uncertain instances to label next.

This way, active learning allows us to strategically select the most informative examples to label, thereby potentially improving the model's performance with fewer labeled instances.

There are several active learning strategies that can be employed based on different criteria for selecting informative samples. Let's discuss a few commonly used strategies and their Python implementations.

Uncertainty sampling

Uncertainty sampling is based on the assumption that instances on which a model is uncertain are more informative and beneficial to label. The idea is to select instances that are close to the decision boundary or have conflicting predictions. By actively acquiring labels for these challenging instances, the model can refine its understanding of the data and improve its performance.

There are several common approaches to uncertainty sampling in active learning:

- **Least confidence**: This method selects instances for which the model has the lowest confidence in its predictions. It focuses on instances where the predicted class probability is closest to 0.5, indicating uncertainty. For instance, in our email example, if the model predicts a 0.52 probability of a particular email being spam and a 0.48 probability of it not being spam, this indicates that the model is uncertain about its prediction. This email would be a prime candidate for labeling under the least confidence method.

- **Margin sampling**: Margin sampling aims to find instances where the model's top two predicted class probabilities are close. It selects instances with the smallest difference between the highest and second-highest probabilities, as these are likely to be near the decision boundary. Let's say we have a model that classifies images of animals. If it predicts an image with probabilities of 0.4 for cat, 0.38 for dog, and 0.22 for bird, the small difference (0.02) between the top two probabilities suggests uncertainty. This image would be chosen for labeling in margin sampling.

- **Entropy**: Entropy-based sampling considers the entropy of the predicted class probabilities. It selects instances with high entropy, indicating a high level of uncertainty in the model's predictions. Using the same animal classification model, if an image gets equal 0.33 probabilities for each class (cat, dog, bird), it shows high uncertainty. This image would be selected for labeling by the entropy method.

Let's implement uncertainty sampling in Python. For this example, we go back to the `credit-g` dataset introduced at the beginning of this chapter. Let's have a look at the features in this dataset to refresh your memory:

	0	1	2	3	4
checking_status	<0	0<=X<200	no checking	<0	<0
duration	6.0	48.0	12.0	42.0	24.0
credit_history	critical/other existing credit	existing paid	critical/other existing credit	existing paid	delayed previously
purpose	radio/tv	radio/tv	education	furniture/equipment	new car
credit_amount	1169.0	5951.0	2096.0	7882.0	4870.0
savings_status	no known savings	<100	<100	<100	<100
employment	>=7	1<=X<4	4<=X<7	4<=X<7	1<=X<4
installment_commitment	4.0	2.0	2.0	2.0	3.0
personal_status	male single	female div/dep/mar	male single	male single	male single
other_parties	none	none	none	guarantor	none
residence_since	4.0	2.0	3.0	4.0	4.0
property_magnitude	real estate	real estate	real estate	life insurance	no known property
age	67.0	22.0	49.0	45.0	53.0
other_payment_plans	none	none	none	none	none
housing	own	own	own	for free	for free
existing_credits	2.0	1.0	1.0	1.0	2.0
job	skilled	skilled	unskilled resident	skilled	skilled
num_dependents	1.0	1.0	2.0	2.0	2.0
own_telephone	yes	none	none	none	none
foreign_worker	yes	yes	yes	yes	yes
target	good	bad	good	good	bad

Figure 6.7 – The features of the credit-g dataset

Under the assumption that the dataset has already been loaded into a `df` DataFrame, we start by preprocessing the dataset by standardizing the numerical features and one-hot encoding categorical features:

```
import pandas as pd
from sklearn.preprocessing import StandardScaler, OneHotEncoder
from sklearn.compose import ColumnTransformer
from sklearn.pipeline import Pipeline
```

```
# Define preprocessor
preprocessor = ColumnTransformer(
    transformers=[
        ('num', StandardScaler(), ['duration', 'credit_amount',
'installment_commitment',
                                'residence_since', 'age',
'existing_credits', 'num_dependents']),
        ('cat', OneHotEncoder(), ['checking_status', 'credit_history',
'purpose',
                                'savings_status', 'employment',
'personal_status', 'other_parties',
                                'property_magnitude', 'other_
payment_plans', 'housing', 'job',
                                'own_telephone', 'foreign_
worker'])])

# Define a mapping from current labels to desired labels
mapping = {'good': 1, 'bad': 0}

# Apply the mapping to the target variable
df['target'] = df['target'].map(mapping)

# Fit and transform the features
features = preprocessor.fit_transform(df.drop('target', axis=1))

# Convert the features to a dataframe
features_df = pd.DataFrame(features)

# Add the target back to the dataframe
df_preprocessed = pd.concat([features_df, df['target'].reset_
index(drop=True)], axis=1)
```

With our new df_preprocessed DataFrame in hand, we can perform uncertainty sampling. We start by importing the necessary libraries and modules, including pandas, NumPy and scikit-learn, for data manipulation and machine learning operations:

```
import pandas as pd
import numpy as np
from sklearn.model_selection import train_test_split
from sklearn.linear_model import LogisticRegression
from sklearn.metrics import accuracy_score
```

We split the df_preprocessed dataset into a small labeled dataset and the remaining unlabeled data. In this example, we randomly select a small portion of 10% as labeled data and leave the rest as unlabeled data:

```
labeled_data, unlabeled_data = train_test_split(df_preprocessed, test_
size=0.9, random_state=42)
```

We define the uncertainty sampling functions—least_confidence, margin_sampling, and entropy_sampling—as discussed earlier.

Here is an explanation of each of these functions:

- least_confidence: This function takes in a 2D array of probabilities with each row representing an instance and each column representing a class. For each instance, it calculates the confidence as 1 – max_probability, where max_probability is the largest predicted probability across all classes. It then sorts the instances by confidence in ascending order. The idea is that instances with lower confidence (i.e., higher uncertainty) are more informative and should be labeled first:

  ```
  def least_confidence(probabilities):
      confidence = 1 - np.max(probabilities, axis=1)
      return np.argsort(confidence)
  ```

- margin_sampling: This function also takes in a 2D array of probabilities. For each instance, it calculates the margin as the difference between the highest and second-highest predicted probabilities. It then sorts the instances by margin in ascending order. The idea is that instances with smaller margins (i.e., closer top-two class probabilities) are more informative and should be labeled first:

  ```
  def margin_sampling(probabilities):
      sorted_probs = np.sort(probabilities, axis=1)
      margin = sorted_probs[:, -1] - sorted_probs[:, -2]
      return np.argsort(margin)
  ```

- entropy_sampling: This function calculates the entropy of the predicted probabilities for each instance. Entropy is a measure of uncertainty or disorder, with higher values indicating greater uncertainty. It then sorts the instances by entropy in ascending order. The idea is that instances with higher entropy (i.e., more uncertainty in the class probabilities) are more informative and should be labeled first:

  ```
  def entropy_sampling(probabilities):
      entropy = -np.sum(probabilities * np.log2(probabilities),
  axis=1)
      return np.argsort(entropy)
  ```

We enter the active learning loop, where we iteratively train a model, select instances for labeling using uncertainty sampling, obtain labels for those instances, and update the labeled dataset.

Firstly, a list named `accuracies` is used to keep track of the accuracy of the model on the labeled data at each iteration.

The active learning loop is then implemented over a specified number of iterations. In each iteration, a logistic regression model is trained on the labeled data, and its accuracy is calculated and stored. The model then makes predictions on the unlabeled data, and the instances about which it is least confident (as determined by the `least_confidence` function) are added to the labeled dataset. These instances are removed from the unlabeled dataset:

```
# Initialize a list to store the accuracy at each iteration
accuracies = []

# Implement the active learning loop.
num_iterations = 5
batch_size = 20

for _ in range(num_iterations):
    model = LogisticRegression()
    model.fit(X_labeled, y_labeled)

    # Calculate and store the accuracy on the labeled data at this
iteration
    accuracies.append(accuracy_score(y_labeled, model.predict(X_
labeled)))

    probabilities = model.predict_proba(X_unlabeled)
    indices = least_confidence(probabilities)[:batch_size]
    X_newly_labeled = X_unlabeled[indices]
    y_newly_labeled = y_unlabeled[indices]
    X_labeled = np.concatenate([X_labeled, X_newly_labeled])
    y_labeled = np.concatenate([y_labeled, y_newly_labeled])
    X_unlabeled = np.delete(X_unlabeled, indices, axis=0)
    y_unlabeled = np.delete(y_unlabeled, indices)
```

The final output is an updated labeled dataset containing additional instances labeled during each iteration of active learning. The process aims to improve model performance by iteratively selecting and labeling the most informative instances from the unlabeled data. In the preceding code, the success and effectiveness of active learning depend on the `least_confidence` custom function and the characteristics of the dataset. The `least_confidence` function is assumed to return indices corresponding to the least confident predictions.

Note that this code uses the `least_confidence` function to perform active learning. To perform the same process using `margin_sampling` or `entropy_sampling` instead of `least_confidence`, you could replace `least_confidence(probabilities)[:batch_size]` with `margin_sampling(probabilities)[:batch_size]` or `entropy_sampling(probabilities)[:batch_size]`.

Let's compare the performance of each of the three active learning functions on the same sample from `credit-g`. We use matplotlib to produce visual representations of the accuracy for each of the three active learning functions. To replicate the output, apply the following code to the outputs of each function:

```
import matplotlib.pyplot as plt
from matplotlib.ticker import MaxNLocator

ax = plt.figure().gca()
ax.xaxis.set_major_locator(MaxNLocator(integer=True))

plt.plot(range(1, num_iterations + 1), accuracies)
plt.xlabel('Iteration')
plt.ylabel('Accuracy')
plt.title('Model accuracy over iterations (least_confidence)')
plt.show()
```

The *least confidence* method achieved a model accuracy of 0.878 after five iterations, with the best performance observed after the first iteration:

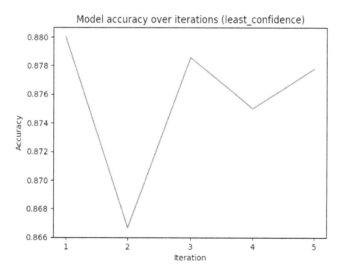

Figure 6.8 – Accuracy of the least_confidence active learning function over five iterations when predicting the target variable on the credit-g dataset

Margin sampling achieved a slightly higher accuracy of 0.9 after two and three iterations:

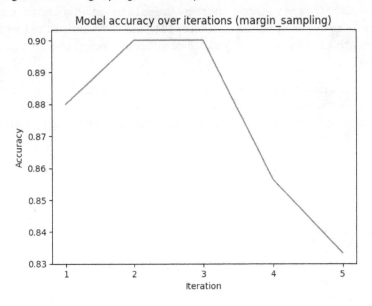

Figure 6.9 – Accuracy of the margin_sampling active learning function over five iterations when predicting the target variable on the credit-g dataset

Lastly, entropy sampling and least confidence achieved identical results. How come?

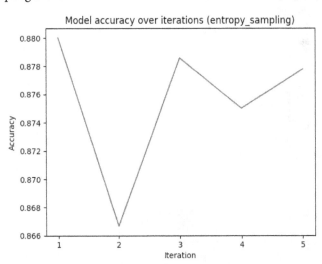

Figure 6.10 – Accuracy of the entropy_sampling active learning function over five iterations when predicting the target variable on the credit-g dataset

These two methods may yield the same results in certain scenarios, especially when working with binary classification problems. Here's why.

The least confidence method considers the class with the highest predicted probability. If the model is very confident about a particular class (i.e., the probability is close to 1), then it's less likely that this instance will be selected for labeling.

Entropy sampling considers the entropy or "disorder" of the predicted probabilities. For binary classification problems, entropy is maximized when the probabilities are both equal (i.e., the model is completely unsure which class to predict). This could coincide with low-confidence predictions.

As a result, both methods will often select the same instances for labeling in the context of binary classification. However, this might not always be the case, especially for multi-class problems.

On the other hand, margin sampling focuses on the difference between the highest and second-highest predicted probabilities. Even slight differences in these probabilities can lead to different instances being selected compared to the other methods.

In the next section, we'll explore **Query by Committee** (**QBC**) for labeling—a method that brings together a group of models to help decide which data points are most important for labeling.

Query by Committee (QBC)

QBC is based on the idea that instances for which the committee of models disagrees or exhibits high uncertainty are the most informative and should be prioritized for labeling. Instead of relying on a single model's prediction, QBC takes advantage of the diversity and collective decision-making of the committee to make informed labeling decisions.

The QBC process typically involves the following steps:

1. **Committee creation**: Create an initial committee of multiple models trained on the available labeled data. Models in a committee can be diverse in terms of their architectures, initializations, or training methodologies.

2. **Instance selection**: Apply the committee of models to the unlabeled instances and obtain predictions. Choose the instances that elicit the most disagreement or uncertainty among the committee members for labeling.

3. **Committee update**: Label the selected instances and add them to the labeled dataset. Re-train or update the committee of models using the expanded labeled dataset.

4. **Repeat**: Iterate the process by returning to *step 2* until a desired performance level is achieved or labeling resources are exhausted.

QBC offers several advantages in active learning for labeling:

- **Model diversity**: QBC utilizes a committee of models, allowing for diverse perspectives and capturing different aspects of the data distribution. This diversity helps identify instances that are challenging or ambiguous, leading to improved labeling decisions.

- **Model confidence estimation**: By observing the disagreement or uncertainty among the committee members, QBC provides an estimate of the models' confidence in their predictions. Instances that lead to disagreement or uncertainty can be considered more informative and valuable for labeling.

- **Labeling efficiency**: QBC aims to prioritize instances that have the greatest impact on the committee's decision. This approach can save labeling efforts by focusing on instances that provide the most relevant information to improve the model's performance.

Let's implement this approach using the `credit-g` dataset in Python. First, we define functions for creating the committee, obtaining committee predictions, and measuring disagreement or uncertainty among the committee members.

The `create_committee(num_models)` function creates a committee of logistic regression models. The number of models in the committee is specified by `num_models`.

The `get_committee_predictions(committee, data)` function gets predictions from each model in the committee for the provided data. It returns an array of prediction probabilities.

The `measure_disagreement(predictions)` function measures the disagreement among the committee's predictions. It calculates the variance of the predictions and returns the mean disagreement:

```python
def create_committee(num_models):
    committee = []
    for _ in range(num_models):
        model = LogisticRegression()
        # Customize and train each model as needed
        committee.append(model)
    return committee

def get_committee_predictions(committee, data):
    predictions = []
    for model in committee:
        preds = model.predict_proba(data)
        predictions.append(preds)
    return np.array(predictions)

def measure_disagreement(predictions):
    disagreement = np.var(predictions, axis=0)
    return np.mean(disagreement, axis=1)
```

We enter the active learning loop, where we iteratively train the committee, measure disagreement or uncertainty, select instances for labeling, obtain labels for those instances, and update the labeled dataset:

```
labeled_dataset = labeled_data.copy()

num_iterations = 5
batch_size = 20
committee = create_committee(num_models=3)

for _ in range(num_iterations):
    for model in committee:
        X_train = labeled_dataset.drop('target', axis=1)
        y_train = labeled_dataset['target']
        model.fit(X_train, y_train)
    X_unlabeled = unlabeled_data.drop('target', axis=1)
    committee_predictions = get_committee_predictions(committee, X_
unlabeled)
    disagreement = measure_disagreement(committee_predictions)
    indices = np.argsort(disagreement)[-batch_size:]
    labeled_instances = unlabeled_data.iloc[indices]
    labels = labeled_instances['Loan_Status']
    labeled_dataset = pd.concat([labeled_dataset, labeled_instances])
    unlabeled_data = unlabeled_data.drop(labeled_instances.index)
```

Here's an explanation of the code. `labeled_dataset = labeled_data.copy()` creates a copy of the initial labeled dataset.

The loop over `num_iterations` represents the number of rounds of semi-supervised learning. In each round, the following steps occur:

1. Each model in the committee is trained on the current labeled dataset.
2. The committee makes predictions on the unlabeled data.
3. The disagreement among the committee's predictions is calculated.
4. The indices of the instances with the highest disagreement are identified. The size of this batch is specified by `batch_size`.
5. These instances are added to the labeled dataset.
6. Finally, these instances are removed from the unlabeled data.

The idea behind this approach is that the instances the models disagree about the most are the ones where the models are most uncertain. By adding these instances to the labeled dataset, the models can learn more from them in the next round of training. This process continues for a specified number of iterations.

Now let's discuss diversity sampling in labeling—a smart technique that focuses on selecting a varied set of data points to ensure a well-rounded and representative labeled dataset.

Diversity sampling

Diversity sampling is based on the principle that selecting instances that cover diverse patterns or regions in the dataset can provide a more comprehensive understanding of the underlying data distribution. By actively seeking diverse instances for labeling, diversity sampling aims to improve model generalization and robustness.

The diversity sampling process typically involves the following steps:

1. **Initial model training**: Train an initial machine learning model using a small, labeled dataset. This model will be used to guide the selection of diverse instances for labeling.

2. **Instance selection**: Apply the trained model to the unlabeled instances and obtain predictions. Calculate a diversity metric to measure the dissimilarity or coverage of each instance with respect to the already labeled instances. Select instances with the highest diversity metric for labeling.

3. **Labeling and model update**: Label the selected instances and add them to the labeled dataset. Retrain or update the machine learning model using the expanded labeled dataset.

4. **Repeat**: Iterate the process by returning to *step 2* until a desired performance level is achieved or labeling resources are exhausted.

Diversity sampling offers several advantages in active learning for labeling:

- **Comprehensive data coverage**: By selecting diverse instances, diversity sampling ensures that the labeled dataset covers a wide range of patterns or regions in the data. This approach helps the model generalize better to unseen instances and improves its ability to handle different scenarios.

- **Exploration of data distribution**: Diversity sampling encourages the exploration of the underlying data distribution by actively seeking instances from different parts of the feature space. This exploration can reveal important insights about the data and improve the model's understanding of complex relationships.

- **Mitigation of bias and overfitting**: Diversity sampling can help mitigate biases and overfitting that may arise from selecting only easy or similar instances for labeling. By diversifying the labeled dataset, diversity sampling reduces the risk of model overconfidence and enhances its robustness.

Let's explore this approach on our preprocessed `credit-g` dataset, using pairwise distances from the `sklearn` library in Python. The `pairwise_distances` function from `sklearn` calculates the distance between each pair of instances in a dataset. In the context of diversity sampling, this function is used to find instances that are most different from each other.

Here's the process:

1. Compute the pairwise distances between all pairs of instances in the unlabeled dataset.

2. Identify the instances that have the greatest distances between them. These are the most diverse instances according to the distance metric used.

3. Select these diverse instances for labeling and add them to the labeled dataset.

The idea is that by actively seeking out diverse instances (those that are farthest apart in terms of the chosen distance metric), you can cover a wider range of patterns in the underlying data distribution. This helps to improve the model's ability to generalize to new data and enhances its robustness.

First, we import the `pairwise_distances` function:

```
from sklearn.metrics.pairwise import pairwise_distances
```

We define functions for calculating diversity and selecting instances with the highest diversity for labeling. We will use pairwise Euclidean distance as the diversity metric:

```
def calculate_diversity(data):
    distance_matrix = pairwise_distances(data, metric='euclidean')
    diversity = np.sum(distance_matrix, axis=1)
    return diversity

def select_diverse_instances(data, num_instances):
    diversity = calculate_diversity(data)
    indices = np.argsort(diversity)[-num_instances:]
    return data.iloc[indices]
```

We enter the active learning loop, where we iteratively calculate diversity, select diverse instances for labeling, obtain labels for those instances, and update the labeled dataset:

```
labeled_dataset = labeled_data.copy()
num_iterations = 5
batch_size = 20
for _ in range(num_iterations):
    X_unlabeled = unlabeled_data.drop('target', axis=1)
    diversity = calculate_diversity(X_unlabeled)
    labeled_instances = select_diverse_instances(unlabeled_data,
batch_size)
    labels = labeled_instances['target']
    labeled_dataset = pd.concat([labeled_dataset, labeled_instances])
    unlabeled_data = unlabeled_data.drop(labeled_instances.index)
```

In the next section, we'll explore transfer learning in labeling—an advanced method that leverages knowledge gained from one task to improve performance on a different but related task.

Transfer learning

Transfer learning involves using knowledge gained from a source task or domain to aid learning. Instead of starting from scratch, transfer learning leverages pre-existing information, such as labeled data or pre-trained models, to bootstrap the learning process and improve the performance of the target task. Transfer learning offers several advantages in the labeling process of machine learning:

- **Reduced labeling effort**: By leveraging pre-existing labeled data, transfer learning reduces the need for the manual labeling of a large amount of data for the target task. It enables the reuse of knowledge from related tasks, domains, or datasets, saving time and effort in acquiring new labels.

- **Improved model performance**: Transfer learning allows the target model to benefit from the knowledge learned by a source model. The source model might have been trained on a large, labeled dataset or a different but related task, providing valuable insights and patterns that can enhance the target model's performance.

- **Adaptability to limited labeled data**: Transfer learning is particularly useful when the target task has limited labeled data. By leveraging labeled data from a source task or domain, transfer learning can help generalize the target model better and mitigate the risk of overfitting on a small, labeled dataset.

Transfer learning can be applied in various ways for labeling in machine learning:

- **Feature extraction**: Utilize pre-trained models as feature extractors. Extract high-level features from pre-trained models and feed them as inputs to a new model that is trained on the target labeled dataset.

- **Fine-tuning pre-trained models**: Use pre-trained models that have been trained on large, labeled datasets, such as models from popular deep learning architectures such as VGG, ResNet, or BERT. Fine-tune these pre-trained models on a smaller labeled dataset specific to the target task.

Let's discuss these in more detail.

Feature extraction

Feature extraction involves using the representations learned by a pre-trained model as input features for a new model. This approach is particularly useful when the pre-trained model has been trained on a large, general-purpose dataset such as ImageNet. Here's an example of using transfer learning for image labeling using the VGG16 model.

> **Note**
> The data used in this example is available from `https://github.com/odegeasslbc/FastGAN-pytorch`.

We first import the necessary libraries:

```
import PIL
import PIL.Image
from tensorflow.keras.applications import VGG16
from tensorflow.keras.preprocessing import image
from tensorflow.keras.applications.vgg16 import preprocess_input,
decode_predictions
import numpy as np
from tensorflow.keras.preprocessing.image import ImageDataGenerator
from tensorflow.keras.models import Sequential
from tensorflow.keras.layers import Dense, Flatten
```

We will use an image of a Golden Retriever for labeling. We can view this image using the `PIL` library:

```
PIL.Image.open('path_to_image_2.jpg')
```

This will display the following image:

Figure 6.11 – The Golden Retriever: man's best friend and our sample image

Let's load the pre-trained VGG16 model from the TensorFlow library and predict the label for this sample image:

```
model = VGG16(weights='imagenet')
img_path = 'path_to_image_2.jpg'
img = image.load_img(img_path, target_size=(224, 224))
x = image.img_to_array(img)
x = np.expand_dims(x, axis=0)
x = preprocess_input(x)
features = model.predict(x)
decoded_predictions = decode_predictions(features, top=5)[0]
for _, label, confidence in decoded_predictions:
    print(label, confidence)
```

This will generate the following output:

```
1/1 [==============================] - 0s 420ms/step
golden_retriever 0.92119014
kuvasz 0.02359655
Tibetan_mastiff 0.014449641
Labrador_retriever 0.011605156
Great_Pyrenees 0.0065631066
```

Figure 6.12 – The VGG16 model's prediction with confidence levels; the model has
labeled the image as golden_retriever with a high level of confidence

The model has correctly predicted the image as `golden_retriever` with `0.92119014` confidence. We now understand how a pre-trained model can be used on a new dataset.

Fine-tuning pre-trained models

Fine-tuning in transfer learning refers to the process of adapting or updating the pre-trained model's parameters to better fit a specific task or dataset of interest. When using transfer learning, the pre-trained model is initially trained on a large-scale dataset, typically from a different but related task or domain. Fine-tuning allows us to take advantage of the knowledge learned by the pre-trained model and customize it for a specific task or dataset.

The fine-tuning process typically involves the following steps:

1. **Pre-trained model initialization**: The pre-trained model, which has already learned useful representations from a source task or dataset, is loaded. The model's parameters are frozen initially, meaning they are not updated during the initial training.

2. **Modification of the model**: Depending on the specific task or dataset, the last few layers or specific parts of the pre-trained model may be modified or replaced. The architecture of the model can be adjusted to match the desired output or accommodate the characteristics of the target task.

3. **Unfreezing and training**: After modifying the model, the previously frozen parameters are unfrozen, allowing them to be updated during training. The model is then trained on the target task-specific dataset, often referred to as the fine-tuning dataset. The weights of the model are updated using backpropagation and gradient-based optimization algorithms to minimize the task-specific loss function.

4. **Training with a lower learning rate**: During fine-tuning, a smaller learning rate is typically used compared to the initial training of the pre-trained model. This smaller learning rate helps to ensure that the previously learned representations are preserved to some extent while allowing the model to adapt to the target task or dataset.

The process of fine-tuning strikes a balance between utilizing the knowledge captured by the pre-trained model and tailoring it to the specifics of the target task. By fine-tuning, the model can learn task-specific patterns and optimize its performance for the new task or dataset. The amount of fine-tuning required may vary depending on the similarity between the source and target tasks or domains. In some cases, only a few training iterations may be sufficient, while in others, more extensive training may be necessary.

Fine-tuning is a crucial step in transfer learning, as it enables the transfer of knowledge from a source task or dataset to a target task, resulting in improved performance and faster convergence on the target task. Here's an example of using transfer learning for image labeling using the VGG16 model. We first import the necessary libraries:

```
from tensorflow.keras.applications import VGG16
from tensorflow.keras.preprocessing.image import ImageDataGenerator
from tensorflow.keras.models import Sequential
from tensorflow.keras.layers import Dense, Flatten
```

We have two classes of images – that of dogs and cats – and therefore we set the number of classes variable to 2. We will also load a pre-trained VGG16 model without the top layers. The image sizes we have here are 256 x 256 x 3:

```
num_classes = 2
base_model = VGG16(weights='imagenet', include_top=False, input_
shape=(256, 256, 3))
```

We now freeze the pre-trained layers and create a new model for fine-tuning. We then compile the model using 'adam' as the optimizer:

```
for layer in base_model.layers:
    layer.trainable = False

model = Sequential()
model.add(base_model)
model.add(Flatten())
model.add(Dense(256, activation='relu'))
model.add(Dense(num_classes, activation='softmax'))

model.compile(optimizer='adam', loss='categorical_crossentropy',
metrics=['accuracy'])
```

To prepare for training, we are configuring data generators for both training and validation datasets. This crucial step involves rescaling pixel values to a range between 0 and 1 using `ImageDataGenerator`. By doing so, we ensure consistent and efficient processing of image data, enhancing the model's ability to learn patterns and features during training:

```
train_data_dir = /path_to_training_data'
validation_data_dir = '/path_to_validation_data'
batch_size = 32

train_datagen = ImageDataGenerator(rescale=1.0/255.0)
validation_datagen = ImageDataGenerator(rescale=1.0/255.0)

train_generator = train_datagen.flow_from_directory(
    train_data_dir,
    target_size=(256, 256),
    batch_size=batch_size,
    class_mode='categorical')

validation_generator = validation_datagen.flow_from_directory(
    validation_data_dir,
    target_size=(256, 256),
    batch_size=batch_size,
    class_mode='categorical')
```

We can now fine-tune the model and save it:

```
epochs = 10
model.fit(
    train_generator,
    steps_per_epoch=train_generator.samples // batch_size,
    epochs=epochs,
    validation_data=validation_generator,
    validation_steps=validation_generator.samples // batch_size)
model.save('fine_tuned_model.h5')
```

With this saved model, we can deploy it for various applications, such as making predictions on new data, integrating it into larger systems, or further fine-tuning similar tasks. The saved model file encapsulates the learned patterns and features from the training process, providing a valuable resource for future use and analysis.

In the next section, we'll delve into the concept of semi-supervised learning in labeling—a sophisticated yet approachable technique that combines the strengths of both labeled and unlabeled data.

Semi-supervised learning

Traditional supervised learning relies on a fully labeled dataset, which can be time-consuming and costly to obtain. Semi-supervised learning, on the other hand, allows us to leverage both labeled and unlabeled data to train models and make predictions. This approach offers a more efficient way to label data and improve model performance.

Semi-supervised learning is particularly useful when labeled data is scarce or expensive to obtain. It allows us to make use of the vast amounts of readily available unlabeled data, which is often abundant in real-world scenarios. By leveraging unlabeled data, semi-supervised learning offers several benefits:

- **Cost-effectiveness**: Semi-supervised learning reduces the reliance on expensive manual labeling efforts. By using unlabeled data, which can be collected at a lower cost, we can significantly reduce the expenses associated with acquiring labeled data.

- **Utilization of large unlabeled datasets**: Unlabeled data is often abundant and easily accessible. Semi-supervised learning enables us to tap into this vast resource, allowing us to train models on much larger datasets compared to fully supervised learning. This can lead to better model generalization and performance.

- **Improved model performance**: By incorporating unlabeled data during training, semi-supervised learning can improve model performance. The unlabeled data provides additional information and helps the model capture the underlying data distribution more accurately. This can lead to better generalization and increased accuracy on unseen data.

There are different approaches within semi-supervised learning that leverage the unlabeled data in different ways. Some common methods include the following:

- **Self-training**: Self-training involves training a model initially on the limited labeled data. Then, the model is used to make predictions on the unlabeled data, and the confident predictions are considered as pseudo-labels for the unlabeled instances. These pseudo-labeled instances are then combined with the labeled data to retrain the model iteratively.

- **Co-training**: Co-training involves training multiple models on different subsets or views of the data. Each model learns from the labeled data and then predicts labels for the unlabeled data. The agreement or disagreement between the models' predictions on the unlabeled data is used to select the most confident instances, which are then labeled and added to the training set for further iterations.

- **Generative models**: Generative models, such as **variational autoencoders** (**VAEs**) or **generative adversarial networks** (**GANs**), can be used in semi-supervised learning. These models learn the underlying data distribution and generate plausible instances. By incorporating the generated instances into the training process, the model can capture more diverse representations and improve its generalization performance.

Let's see a simple implementation in Python of this labeling approach. First, we import the necessary libraries:

```
import numpy as np
import pandas as pd
from sklearn.model_selection import train_test_split
from sklearn.preprocessing import StandardScaler
from sklearn.linear_model import LogisticRegression
from sklearn.metrics import accuracy_score
from sklearn.semi_supervised import LabelPropagation
```

We utilize the preprocessed `credit-g` dataset from previous examples and split it into labeled and unlabeled subsets. This example assumes that you are using the `df_preprocessed` DataFrame we created in the *Uncertainty sampling* section:

```
X = df_preprocessed.drop('target', axis=1)
y = df_preprocessed['target']

# Split the dataset into labeled and unlabeled
X_train, X_test, y_train, y_test = train_test_split(X, y, test_
size=0.2, random_state=42)

labeled_percentage = 0.1  # Percentage of labeled data
X_train_labeled, X_train_unlabeled, y_train_labeled, _ = train_test_
split(
    X_train, y_train, test_size=1 - labeled_percentage, random_
state=42)
```

Then we train a supervised machine learning model using the labeled data. In this example, we will use logistic regression as the supervised model:

```
supervised_model = LogisticRegression()
supervised_model.fit(X_train_labeled, y_train_labeled)
```

We then apply the trained supervised model to predict labels for the unlabeled data. The predicted labels are considered as pseudo-labels for the unlabeled instances:

```
# Predict labels for the unlabeled data
pseudo_labels = supervised_model.predict(X_train_unlabeled)
```

Now concatenate the labeled data (`X_labeled`) with the pseudo-labeled data (`X_unlabeled`) to create the combined feature dataset (`X_combined`). Concatenate the corresponding labels (`y_labeled` and `pseudo_labels`) to create the combined label dataset (`y_combined`):

```
# Concatenate the labeled data with the pseudo-labeled data
```

```
X_combined = np.concatenate((X_labeled, X_unlabeled))
y_combined = np.concatenate((y_labeled, pseudo_labels))
```

Next, train a semi-supervised machine learning model using the combined feature dataset (X_combined) and label dataset (y_combined). In this example, we will use LabelPropagation as the semi-supervised model:

```
semi_supervised_model = LabelPropagation()
semi_supervised_model.fit(X_combined, y_combined)
```

Use the trained semi-supervised model to make predictions on the test set and calculate the accuracy:

```
y_pred = semi_supervised_model.predict(X_test)

accuracy = accuracy_score(y_test, y_pred)
print(f'Test accuracy: {accuracy:.3f}')
```

The print statement outputs the resulting accuracy score, which, in this case, is 0.635.

After training our semi_supervised_model using LabelPropagation, the resulting model has effectively learned from both labeled and unlabeled data. The predictions on the test set (y_pred) showcase the model's ability to generalize and infer labels for previously unseen instances. This output serves as a valuable demonstration of how semi-supervised learning techniques, leveraging both labeled and unlabeled data, can contribute to robust and accurate predictions in real-world scenarios.

Summary

In this chapter, we explored various programmatic labeling techniques in machine learning. Labeling data is essential for training effective models, and manual labeling can be time-consuming and expensive. Programmatic labeling offers automated ways to assign meaningful categories or classes to instances of data. We discussed a range of techniques, including pattern matching, DB lookup, Boolean flags, weak supervision, semi-weak supervision, slicing functions, active learning, transfer learning, and semi-supervised learning.

Each technique offers unique benefits and considerations based on the nature of the data and the specific labeling requirements. By leveraging these techniques, practitioners can streamline the labeling process, reduce manual effort, and train effective models using large amounts of labeled or weakly labeled data. Understanding and utilizing programmatic labeling techniques are crucial for building robust and scalable machine learning systems.

In the next chapter, we'll explore the role of synthetic data in data-centric machine learning.

7

Using Synthetic Data in Data-Centric Machine Learning

In previous chapters, we discussed various approaches to improving data quality for machine learning purposes through better collection and labeling.

Although human labelers, data ownership, and technical data quality improvement practices are critical to data centricity, there are limits to the kind of labeling and data creation that can be performed by individuals or through empirical observation.

Synthetic data has the potential to fill in these gaps and produce comprehensive training data at a fraction of the cost and time of other approaches.

This chapter provides an introduction to synthetic data generation. We will cover the following main topics:

- What synthetic data is and why it matters for data centricity
- How synthetic data is being used to generate better models
- Common techniques used to generate synthetic data
- The risks and challenges with synthetic data use

Let's start by defining what synthetic data is.

Understanding synthetic data

Synthetic data is artificially created data that, if done right, contains all the characteristics of production data.

The reason it's called synthetic data is that it doesn't have a physical existence – that is, it doesn't come from real-life observations or experiments that we create to gather data that we subsequently use to run analysis or build machine learning models on.

A foundational principle of machine learning is that you need a lot of data, ranging from thousands to billions of observations. The amount you need depends on your model.

As we have outlined many times already, when the required volume of data is difficult to come by, one approach is to improve the signal in your data to make it possible to produce accurate and relevant outputs, even on smaller datasets.

Another option is to create synthetic data to cover the gaps. A major benefit of synthetic data is its scalability. Real training data is collected linearly, one example at a time, which can be both time-consuming and expensive.

In contrast, synthetic data can be generated in very large quantities in a relatively short time and typically at a lower cost. As an example, a training image that may cost $5 if it's obtained from a labeling service might cost $0.05 if it's produced artificially[1].

Synthetic data is touted as the answer to many challenges in the development of more powerful machine learning and AI solutions. From solving privacy issues to inexpensively generating rare but important observations for your modeling and training data, synthetic data can fill the gaps where real-world data falls short.

According to Gartner predictions[2], 60% of the data used in AI and analytics projects will be synthetically generated rather than gathered through real-world observations by 2024.

Traditionally, the use of data for analytical purposes has been driven by the data we have available and its limitations. We might imagine the perfect data solution, but often, the depth, breadth, reliability, and privacy constraints of a dataset limit what we can do in reality. To a large extent, this is what synthetic data aims to fix.

There are different ways to create synthetic data, and to some extent, the technical creation of the data is the least complex part. Validating whether a synthetic dataset is a relevant reflection of potential real-world scenarios and defending against unwanted bias can be time-consuming and challenging.

If you choose to use synthetic data for your next project, the first important question is always, "*What are you going to use the data for?*" The answer to this question determines your data needs, which, in turn, will highlight your data gaps.

Let's take a closer look at the typical reasons for using synthetic data and explore some common use cases.

The use case for synthetic data

The reasons for using synthetic data generally fall into the following four categories:

- **Availability**: Synthetic data creation is used to compensate for the lack of data in a domain. It may be that we have imbalanced classes in a dataset compared to the real-life distribution, so to make those classes balanced, we create synthetic data to compensate.

- **Cost**: It can be very costly and time-consuming to collect certain types of data, in which case it can be useful to generate synthetic data to reduce the time and cost spent on a project.

- **Risk management**: In some cases, synthetic data can also be used to lower the risk of human or financial damage. An example of this is flight simulators, which are used to train new and experienced pilots in all sorts of situations. Training pilots in a simulated environment allows us to safely and knowingly introduce rare events that would be hard to create in a natural environment without unacceptable risk.

- **Security and legal compliance**: The data you need may already exist but it is unsafe or illegal to use it for machine learning purposes. For example, some regulations, such as Europe's **General Data Protection Regulation** (**GDPR**), forbid the use of certain kinds of data without clear consent from the underlying individual. Alternatively, it might just be too slow and cumbersome to get signoff in your organization.

Here are some examples of common and potential use cases for synthetic data:

- Computer vision and image and video processing

- Natural language processing

- Privacy preservation

- Correcting bias (discussed in *Chapter 8, Techniques for Identifying and Removing Bias*)

- Improving data quality or gaps (more cheaply)

- Increasing modeling data volumes for rare events (discussed in *Chapter 9, Dealing with Edge Cases and Rare Events in Machine Learning*)

- Simulation

We will explore some of these topics in this chapter to illustrate how synthetic data can be used as part of your model development strategy.

To set the scene, let's look at an example of just how powerful synthetic data can be in the right setting, courtesy of the world's leading computer games software development company, Unity Technologies. By way of background, the Unity platform was used to create 72% of the top 1,000 mobile phone games and 50% of all computer games across mobile, PC, and consoles in 2021[3].

The users of Unity's technology have improved object recognition rates from 70% to almost 100% simply by augmenting real-world data with synthetic data. Synthetic data adds a lot more variety and many more scenarios to the training data, which enables objects to be recognized from many angles[4].

Unity's Vice President of AI and machine learning, David Lange, says the following:

"We're using the Unity engine to recreate three-dimensional worlds with objects in there. Then, we can generate synthetic images that look very much like what they would look like in the real world, perfectly labeled.

"Real-world data is really just a snapshot of the situation. What you can do with the synthetic data is augment that real world with special use cases, special situations, special events. You can improve the diversity of your data by adding synthetic data to your dataset.

"We can create improbable situations because it's not going to cost us anything in milliseconds, rather than trying to stage them in reality. The ease with which you can create all these scenarios is driving the use of synthetic data."

David Lange shares the view of Gartner in that synthetic data is going to be the predominant raw material for machine learning in the future:

"I believe that the vast majority of training data will be synthetic. You have to have the real world as a baseline, but synthetic data eliminates privacy concerns because there are no real people involved. You can eliminate bias. You can do your data analytics and ensure that your data represents the real world in a very even way, better than the real world does."

Take note of the benefits of synthetic data that David Lange mentions:

- Objects can be perfectly labeled, thereby avoiding the labeling ambiguities discussed in previous chapters

- The diversity of the dataset can be increased substantially to cover slight variations in probable scenarios, as well as rare events and edge cases

- Datasets can be scaled quickly and cheaply

- Bias and privacy concerns can be reduced because data is cleaner and depersonalized

Let's dig deeper into the various uses of synthetic data to understand the possibilities, benefits, risks, and constraints associated with it.

Synthetic data for computer vision and image and video processing

At the time of writing, the most prevalent use of synthetic data is in computer vision problems. This is because we can often create this type of data with limited risk, while outliers (often rare but impactful events) can be particularly hard to get hold of in image data.

A common challenge in computer vision (and most other machine learning problems for that matter) is that real-world data typically contains a large proportion of observations describing the most probable scenarios, and very few or no examples of rare events.

At the same time, real-world data can be difficult, expensive, or outright dangerous to collect. As an example, autonomous vehicle models can't be trained to avoid car crashes by putting real cars into dangerous situations. Instead, these crashes and other rare but significant events must be simulated.

A common problem for image classification algorithms is recognizing familiar objects in slightly unfamiliar positions or environments. Because machine learning algorithms don't reason by logic or abstraction, even models that perform very well on both training and test datasets will often fail to generalize to out-of-distribution observations. This is true whether these observations are introduced as an adversarial test of model performance or occur naturally.

Figure 7.1 provides a simplified example of this phenomenon. A square that is rotated 45 degrees may be interpreted by some – humans and algorithms alike – as a diamond, and not simply a tilted square with the same dimensions as a square positioned on its side:

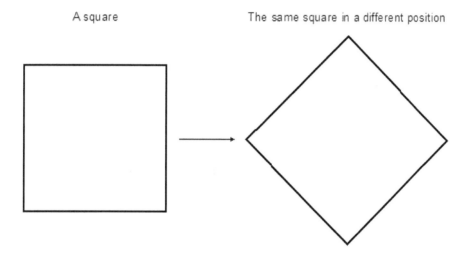

Figure 7.1 – These two squares are either identical or different,
depending on the rules we use to interpret them

The implications of this bias are often substantial. In an analysis of deep neural networks' performance on images from the ImageNet dataset, Alcorn et al. (2019)[5] describe how the common image classifiers *Google Inception-v3*, *AlexNet*, and *ResNet-50* can easily be fooled by slight changes to the positioning of an object within an image.

In *Figure 7.2*, the images in column (d) are real photographs collected from the internet, whereas columns (a) to (c) are out-of-distribution images of the same objects in unusual positions. The main object within the image is flipped and rotated, but the background is kept constant:

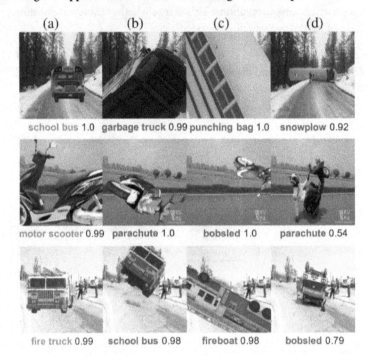

Figure 7.2 – Deep neural networks can easily be fooled when familiar objects are in uncommon positions. Column (d) represents real-life images, while columns (a) to (c) are synthetic

The authors then used *Inception-v3* to classify these images, with the resulting label and confidence score depicted under each image. In these examples, the algorithm was able to classify with a high degree of accuracy and confidence when objects were in a commonly observed position in real-life scenarios. However, the algorithm misclassified images with a high degree of confidence when objects were being flipped, rotated, or moved very close.

Not only did the algorithm misclassify objects, but it also confidently mislabeled them as objects they are not. A rolling bus and a punching bag are like chalk and cheese, and an overturned scooter is nowhere close to looking like a parachute.

Being able to recognize familiar objects in unfamiliar positions is especially critical when it comes to observing and classifying moving objects. In the example of self-driving cars, algorithmic misinterpretations introduce novel and unexpected events into the traffic environment, even though these vehicles are statistically safer than human drivers[6].

The following image, which was captured from a traffic camera in Taiwan, shows an example of this issue. A truck has overturned on a busy highway, and an autonomous Tesla sedan doesn't recognize the truck as an obstacle in the way and crashes into the truck at high speed:

Figure 7.3 – An autonomous vehicle crashes into an overturned truck at high speed.
Algorithms can be trained to handle novel situations like this using synthetic data

In this very unlikely but highly dangerous scenario, the car's algorithms are not equipped to correctly assess the statistical probability that the object in front of it is blocking the road.

This is a scenario where synthetic data proves highly valuable. As we have just learned, even best-in-class computer vision algorithms have a high degree of sensitivity to variations in the position of common objects. Therefore, objects must be introduced in various positions and lighting conditions in the training data to cover all possible combinations, especially highly improbable ones.

When we use machine learning to figure out what's happening in an image, we are extracting the concave and convex curves within the image, also known as the features within a deep neural network.

To create these curves from a synthetic data perspective, you would simply be recreating those formations within images. This would typically involve flipping, rotating, zooming, cropping, making light changes, and resizing images to create slight variations on the same scenario.

In the example of the Tesla accident, we might create images with the truck driving as normal, rolled on its side, rolled on its back, driving in the wrong direction, in the dark, partially covered by other objects, and so on. These scenarios are hard to get a hold of in real-life imagery, yet they're very important to be able to deal with when the situation arises.

Generating synthetic data using generative adversarial networks (GANs)

GANs are common tools for generating synthetic image data with properties similar to real-world data. GANs were invented by computer scientist Ian Goodfellow in 2014 and have since led to an explosion in generative models that can create all sorts of content, including text, art, video, and images.

GANs are a form of unsupervised learning where a generator model is pitted against a discriminator model (hence the "adversarial" aspect). Both models are neural networks that compete against each other to turn the exercise into a "pseudo-supervised" learning problem.

The generator identifies patterns in the original dataset that are then used to generate synthetic output that could have conceivably existed in the input data. The generated examples become negative training samples for the discriminator model.

It is the discriminator's job to classify the newly generated data as fake or real. This zero-sum contest continues until the discriminator picks *fake* observations as *real* close to 50% of the time.

Mathematically, the training process for a GAN can be thought of as minimizing a loss function that measures the difference between the generated examples and the real examples. This loss function is typically a combination of two terms: one that measures how well the generative model can produce examples that are similar to the real examples, and one that measures how well the discriminative model can distinguish between real and generated examples.

By training the two parts of the GAN in this way, the generative model can learn the patterns and features of the real examples in the training dataset, and then use that information to generate new examples that are similar to the real ones. This allows GANs to be used for a wide range of applications, including image generation, text generation, and many others.

Figure 7.4 provides a conceptual illustration of how GANs iterate through a large number of mini-contests to arrive at a model that can generate very realistic outputs:

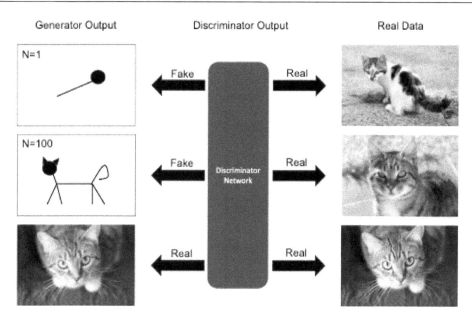

Figure 7.4 – A conceptual illustration of how a GAN works

The generator starts with some very basic presentations of the desired output but gets better at fooling the discriminator as it iterates through many examples. Training is completed when the discriminator struggles to recognize real from fake.

The progressive growth of GANs

As you can imagine, the first few examples that are created by the generator will be relatively easy for the discriminator to pick. There is a lot for the generator model to learn and it can be challenging to make the GAN follow a learning path we would like it to take.

GANs are inherently unstable models, especially when it comes to generating complex structures, such as images. To fool the discriminator, the generator must pick up the small details and larger structures of an image, which can be difficult on high-resolution images. If the generator can't do this, it will get stuck in a no-man's land of never fooling the discriminator.

Another challenge is that large images require lots of computer memory. As a result, the batch size (the number of images used to update model weights each training iteration) must often be reduced to make sure the images will fit into memory. Again, this makes the training process less stable.

A solution to these problems is to progressively increase the detail and complexity of the model's input and output. Progressive growing was first proposed by NVIDIA researchers Karras et al.[7] in 2017, and is a technique for training GANs that allows the model to gradually increase the resolution of the generated images over many iterations.

Under progressive growing, the model is trained using a step-wise approach. First, the generator and discriminator models are trained with low-resolution images and seek to improve image quality by changing their parameters to optimize loss functions. Then, the resolution of the generated images is increased while fine-tuning occurs based on the understanding gathered from the initial training stage until the desired resolution is reached. In other words, the model learns in steps, rather than all at once.

Karras et al. propose training both the generator and discriminator with a batch of low-resolution images of 4x4 pixels. Then, a new sampling layer is used to gradually grow the image complexity to 8x8, using nearest neighbor interpolation.

New network layers are introduced gradually to create minimal disruption between resolution layers. This approach allows for the smooth and seamless integration of newer components into the existing infrastructure.

The gradual phasing in of a new block of layers is done by adding higher-resolution inputs to the existing input or output layer. The relative influence of the new outputs is controlled using a weighting, α, where the weight of the original output is $1 - \alpha$. As α increases, the old layer is gradually faded out, while the new layer takes over.

This process continues until the desired image resolution is reached. *Figure 7.5*, from Kerras et al., highlights the process of progressively growing from 4x4 to 1,024x1,024-pixel images:

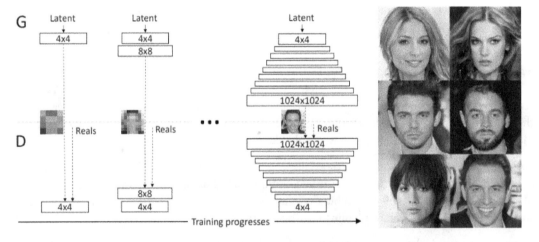

Figure 7.5 – Visualization of the progressive growth of a GAN from Kerras et al. (2017)

This means that the model can first learn about the big picture of the image, and then focus on smaller details. This typically yields better results than trying to learn everything at once. By leveraging this approach, GANs can grasp the essential architecture and characteristics of low-resolution datasets, thus creating higher-quality images with greater precision.

Achieving greater accuracy with StyleGANs

The research team at NVIDIA built on their progressive GAN architecture to introduce the first StyleGAN in December 2018[8]. Since then, StyleGAN-2 and StyleGAN-3 architectures have been released. These incremental upgrades resolved some systemic issues in the output from the original StyleGAN, but are otherwise similar in structure.

The primary innovation of StyleGANs is the ability to control the *style* of the output created by the generator model. The new architecture allows the generator to automatically separate broader features from stochastic/random features in an image. Examples of broad features are a person's pose and identity; hair and freckles are considered stochastic.

Before the introduction of StyleGANs, the inner workings of image generators were partially a mystery and therefore hard to control. With no effective method to compare different images produced by various models and a limited understanding of how features originated, the original GAN generators were black boxes.

Let's take a look at a comparison between the **progressive GAN** (**ProGAN**) and StyleGAN architectures in *Figure 7.6* to understand why StyleGAN has been so successful in generating highly realistic synthetic images:

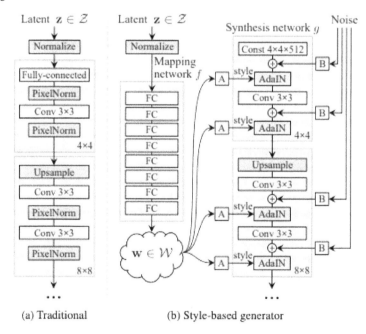

Figure 7.6 – Comparison between ProGAN (a) and StyleGAN (b) from Kerras et al. (2018)

While ProGANs use a progressive training methodology to grow the resolution of generated images layer by layer, StyleGAN gives users more control over the generated images via the use of a Mapping Network and Synthesis Network.

StyleGAN's Mapping Network is a type of neural network that maps a low-dimensional vector, z, to an intermediate latent space, w. This is known as disentangled representation learning. By disentangling the features in these two spaces, users can more easily control the different aspects of the generated image, such as its physiology, hairstyle, or clothing. In simpler terms, it allows for separated control over high-level features of the image rather than specific pixel values.

The Synthesis Network is a deep convolutional neural network that works by receiving style vectors, w, as input and returns an output image. In synthesizing a realistic image, features are pulled from the feature vector and applied to the image layer by layer, beginning with the lowest layer and progressing one layer at a time to higher resolutions.

The Synthesis Network interacts with a learned **adaptive instance normalization** (**AdaIN**) module that rescales image features to increase diversity in image outputs. The module accepts the feature vector and a style vector as inputs and adjusts image features' scaling and bias by subtracting the feature map's mean and dividing it by the standard deviation. As a result, StyleGAN can produce highly detailed images by focusing on specific features such as hairstyle or eye color.

Understanding the challenges of GANs

Although GANs are a wonderful addition to the machine learning toolbox, they are not without their challenges.

GANs are based on zero-sum game theory. Essentially, if one player triumphs, then the other will be defeated. This kind of situation is also known as minimax: your opponent looks to maximize their output while you seek to minimize it. The theory behind GANs states that the game between the generator and discriminator models will continue until a Nash Equilibrium is reached.

The Nash Equilibrium is an important solution concept within economics, politics, and evolutionary biology that can be seen in a wide variety of real-world scenarios. A Nash Equilibrium is a situation where no player has an incentive to do something different than what they are already doing. This is because they have considered what everyone else is doing and they think that their current strategy is the best possible option.

In such situations, all players are said to be at equilibrium as they have no incentive to change their behavior because any changes made by one player will likely lead to a worse outcome for that particular player. Therefore, it is in each individual's best interest not to make any sudden changes in this type of equilibrium situation.

For example, consider a scenario where competing firms are trying to set prices for their products or services. If each firm sets its price too high, it may lose customers to its competitors. However, if each firm sets its price too low, it will not be able to cover its costs and make a profit. Thus, the Nash Equilibrium for this situation is for each firm to set their prices at a level that is low enough to deter customers from buying from their competitors without being so low that they are unable to make a profit.

Although this theory can work, it is often difficult to achieve in practice. There is no guarantee that cost functions will *converge* and find a Nash Equilibrium. In this situation, the game continues indefinitely.

Moreover, when one agent outmatches the other in terms of power and efficacy, the learning signal for their counterpart becomes useless; consequently, no knowledge is gained by either side. The most common scenario is that the discriminator becomes so good at picking the generator's faults that the generator never learns how to advance in the game.

One of the main challenges with GANs is called mode collapse. Like any other statistical model, GANs tend to find the easiest way through the underlying data, which can lead to the overrepresentation of modal observations.

Mode collapse is another common issue that occurs when the generator produces an especially plausible output, which causes it to only produce that output. Once this happens, the discriminator is more likely to fall into a local minimum, unable to find a better output that it deems valid.

Consequently, the generator is driven to tailor its outputs toward the criteria used by this static discriminator rather than attempting to create realistic or dynamic outputs. As such, generators tend to "over-optimize" for a particular outcome, as determined by their single discriminator.

An example of mode collapse is illustrated in *Figure 7.7* and is from Metz et al. (2017)[9]. In this example, the researchers used the MNIST dataset of handwritten digits to train two different GANs. The MNIST dataset contains 10 different modes that represent digits 0-9.

The top four quadrants of numbers have been successfully generated (using an unrolled GAN training method) to look like real handwritten digits with a representation of all possible digits. The bottom four quadrants, on the other hand (generated using the original GAN architecture), have suffered from mode collapse early on in the process and produce only representations of the number 6:

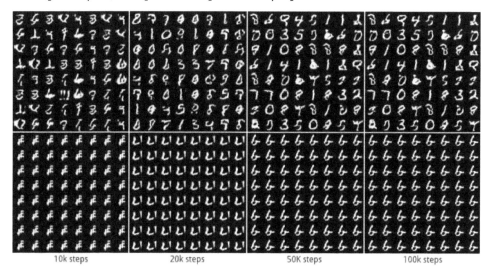

Figure 7.7 – Two different GAN architectures trained on the MNIST dataset from Metz et al. (2017)

Since the development of the original GAN architecture in 2014, several new GAN variants have been introduced to deal with mode collapse. As a result, this is now a less common issue, but it's something to always watch out for.

The mode collapse in the MNIST example is relatively easy to spot, but it is important to note that GANs may potentially preserve and exacerbate existing biases in more subtle ways.

In a 2021 study (Jain et at, 2021)[10], researchers from Arizona State University and Rensselaer Polytechnic Institute assessed the performance of GANs in generating synthetic facial data. The researchers wanted to test whether GANs would exacerbate the modal facial characteristics from a (naturally) biased input dataset, thus increasing the most common features in the synthetic output.

Two model architectures were compared: **deep convolutional GAN** (**DCGAN**) and ProGAN. The experiment involved a dataset of 17,245 images of engineering professors from US universities, of which 80% were classified as male and 76% were classified as white. The experiment used human classifiers (Turkers) to classify the faces in the original dataset and those produced by the GANs.

Figure 7.8 shows the outcomes of two of the GANs tested against the original dataset. The DCGAN model resulted in heavily biased data generation, with a large overrepresentation of males and whites in the synthetic images. While the ProGAN model carried less of a bias, it still wasn't a reasonable representation of the features in the original dataset.

Both DCGAN and ProGAN penalized images with mostly feminine features. DCGAN generated the most biased output, reducing the percentage of feminine faces from 20% to 6.67%.

In comparison to the 24% non-white faces in the original dataset, both DCGAN and ProGAN reduced that rate significantly – 1.33%% for DCGAN and 11.33% for ProGAN, respectively:

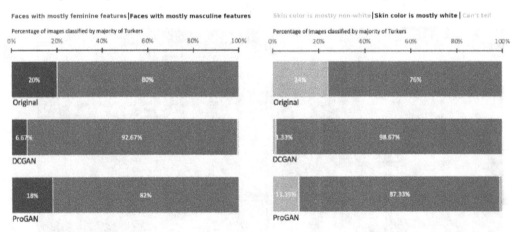

Figure 7.8 – Comparison of gender and skin color distributions from synthetic facial images generated by GANs versus the original dataset. Source: Jain et al., 2021

In other words, GANs can produce highly realistic synthetic datasets that may still be a biased representation of the latent features in the original dataset. Ironically, this partial mode collapse may produce more realistic images because the GAN has specialized to perform well for certain dimensions.

Following our principles of data-centric machine learning, it is important to identify modal bias across all dimensions of a synthetic dataset to ensure it is useful and ethically appropriate for your intended purpose. It is typically a process of trial and error to empirically validate and remove bias in GAN outputs.

Finally, a notable weakness of GANs is that generating high-quality outputs requires a lot of computational resources, and without these, the images that are produced may appear blurry or unrealistic. Furthermore, without an experienced person choosing the appropriate directions to use for image generation, using GANs may not produce the desired outcome.

Exploring image augmentation with a practical example

Despite the complexity involved in synthetic image data generation, we want to finish this section with a practical example that gets you inspired to apply these techniques in your work.

In this section, we will cover data augmentation, a mechanism for generating synthetic data for image data. We will use a pre-trained Xception model that was trained on ImageNet data and fine-tune it to accommodate clothing examples. We will achieve this by applying transfer learning to fine-tune the clothing examples, and then generate synthetic data to enhance its performance. With transfer learning, we can freeze the pre-trained layers of the network and only train the new layers by updating the final output. This helps the model to quickly adapt to the new dataset with less training time.

By applying transfer learning and image augmentation techniques on top of a pre-trained Xception model, we can generate synthetic data that can improve the performance of the model on new data. This approach is widely used in various applications, including image classification, object detection, and segmentation.

To start, we need to load the pre-trained Xception model using the TensorFlow library. We can do this by simply importing the Xception model from the TensorFlow module and setting the `include_top` parameter to `False` to exclude the top layer of the model.

Next, we must add a custom classifier on top of the pre-trained Xception model. We can achieve this by adding a few dense layers and a final output layer with the number of classes equal to the number of clothing categories we want to classify.

To further improve the model's performance, we will apply image augmentation using the out-of-the-box TensorFlow features. This can include rotation, zooming, flipping, and other techniques to create variations in the training data and make the model more robust.

Finally, we will train the model using the augmented training data and evaluate its performance on the validation set. We can fine-tune the hyperparameters of the model and the augmentation techniques to achieve better accuracy.

> **Note**
>
> This example has been adapted from *Chapter 7* of *Machine Learning Bookcamp* (`https://livebook.manning.com/book/machine-learning-bookcamp/chapter-7/`) and has also been made available by the author of the book through a course run by the author's company, DataTalks Club (`https://github.com/alexeygrigorev/mlbookcamp-code`).

To start, we will import all the necessary libraries:

```
import tensorflow as tf
from tensorflow.keras.applications.xception import Xception,
preprocess_input, decode_predictions
from tensorflow.keras.preprocessing.image import load_img,
ImageDataGenerator
import numpy as np
from tensorflow import keras
import matplotlib.pyplot as plt
2024-01-29 17:34:32.614910: I tensorflow/core/util/util.cc:169] oneDNN
custom operations are on. You may see slightly different numerical
results due to floating-point round-off errors from different
computation orders. To turn them off, set the environment variable
`TF_ENABLE_ONEDNN_OPTS=0`.
```

Next, we will download the data by cloning the dataset from its respective GitHub repository:

```
!git clone git@github.com:alexeygrigorev/clothing-dataset-small.git
```

Next, we will load an image of pants that we have downloaded from the clothing dataset repository. To load the image, we will use the `load_img` function:

```
path = './clothing-dataset-small/train/pants/12bfe0f0-accc-4539-ab51-
53f63534938e.jpg'
load_img(path)
```

This will display the following output:

Figure 7.9 – A pair of pants as output from the load_img function

Next, we will represent the image as a NumPy array as this is the format in which the model expects the image. We also need to ensure that all images are passed to the model with the same dimensions, so we will choose a standard of (299, 299, 3), which means 299 pixels top to bottom, 299 pixels left to right, and three color channels, which are red, green, and blue. Each pixel will be represented three times with values from 0-255 for each color channel.

Now, we must load the preceding image with a target size of (299,299):

```
img = load_img(path=path, target_size=(299,299))
img_input = np.array(img)
img_input.shape

(299, 299, 3)
```

Let's see the image with our new dimension:

```
img
```

The output is as follows:

Figure 7.10 – Our image of a pair of pants, resized to be 299x299 pixels

Now that we have ensured the dimensions, we will load the pretrained Xception model and specify the input format, which is the same as the dimensions specified for the preceding image. Once we've loaded the model, we will score the image and check the prediction:

```
model = Xception(weights='imagenet', input_shape=(299,299,3))
```

The model expects the list of images in a particular format; hence, we will pass the previous image as a list and convert it into a NumPy array. Then, we will preprocess the image using `preprocess_input` and use and then score the input. The prediction will be stored in `pred`:

```
img_preprocessed = preprocess_input(np.array([img_input]))
pred = model.predict(img_preprocessed)
pred.shape

1/1 [==============================] - 3s 3s/step

(1, 1000)
```

The `preprocess_input` function is required to preprocess the image data required by the model since when the Xception model was trained, the input values were transformed from 0-255 and scaled to have values between -1 and 1. This is important because the distribution of color scales may affect the prediction. Imagine that red color scales were between 0-100, while blue color scales were between 200-300; this may have led to an unstable model. Hence, scaling is important. Without the correct preprocessing, the predictions won't make sense.

Next, we will decode these predictions with a convenience function. The model will provide a probability of the top five labels out of 1,000 different classes since the Xception model was trained to predict 1,000 labels. We don't believe the target labels consist of clothing examples, so after the next few steps, we will move on to transfer learning. To decode the predictions, we will use the `decode_predictions` function. `decode_predictions` is a convenience function that provides predictions in such a format that they can be easily understood:

```
decode_predictions(pred)

[[('n03594734', 'jean', 0.651147),
  ('n04371430', 'swimming_trunks', 0.22369406),
  ('n03710637', 'maillot', 0.004711655),
  ('n04525038', 'velvet', 0.0038891942),
  ('n03595614', 'jersey', 0.003085624)]]
```

It's quite clear that the image of the pants is closest to the `jean` label and then the `swimming_trunks` label, as per the labels used in the Xception model. However, in the training data, there is no `jean` or `swimming_trunks` label. Next, we will extract the training data and make sure it is preprocessed before we pass it for training.

For this, we will use transfer learning and leverage the entire training data. We will use `ImageDataGenerator` to help process the input data that the model requires and then create training and validation datasets. The training data will consist of 3,041 images, while the validation data will consist of 341 images. For preprocessing, we will use 150x150 pixels instead of 299x299 pixels to reduce the training time:

```
train_gen = ImageDataGenerator(preprocessing_function=preprocess_
input)
```

Next, we will use the `flow_from_directory` property to process the entire training and validation dataset. We will use `seed=42` to ensure data is passed to the network, it's passed with the same randomization, and the result at each training layer is reproducible. For the validation data, we will utilize `shuffle=False` to ensure that there is no randomization at each training step but data is passed sequentially:

```
train_ds = train_gen.flow_from_directory(directory='./clothing-
dataset-small/train/', target_size=(150,150), batch_size=32, seed=42)

validation_ds = train_gen.flow_from_directory(directory="./clothing-
dataset-small/validation/", target_size=(150,150), batch_size=32,
shuffle=False)

Found 3081 images belonging to 10 classes.
Found 341 images belonging to 10 classes.
```

Now, we can view the target classes of our dataset:

```
train_ds.class_indices

{'dress': 0,
 'hat': 1,
 'longsleeve': 2,
 'outwear': 3,
 'pants': 4,
 'shirt': 5,
 'shoes': 6,
 'shorts': 7,
 'skirt': 8,
 't-shirt': 9}
```

Next, we will apply transfer learning. To do so, we must first extract the base layer – in other words, we must extract the convolutional layer of the Xception model and ensure it is frozen. This will ensure we get access to the feature map of the image data of 1 million+ images. To do so, we will use the `include_top=False` parameter, which will ignore the dense layers and return the bottom layer of the convolutional neural network. Next, we will build a custom dense layer with the 10 class labels highlighted previously and train the dense layer for our use case:

```
base_cnn_model = Xception(weights='imagenet', include_top=False,
input_shape=(150,150,3))
base_cnn_model.trainable = False
```

Next, we will build the architecture of the dense layer and combine it with the base layer. First, we will define the input standard such that the base model can provide the vector that's suitable for input of the same standard. We will use (150,150,3) so that the model can be trained faster as more pixels can slow down the training process. Next, we will transform the vector from a base layer into a two-dimensional array. For that, we will use a method called pooling, which can help reduce the spatial size of feature maps but still retain the key information about the base layer. After that, we will create a dense layer where we specify the number of outputs, which is 10 in this case, and apply softmax activation to return a probability.

At this point, we can define the loss function, utilize the Adam optimizer with a learning rate of 0.005, and choose accuracy as a metric:

```
inputs = keras.Input(shape=(150,150,3))
base = base_cnn_model(inputs)
vectors = keras.layers.GlobalAveragePooling2D()(base)
inner = keras.layers.Dense(100, activation='relu')(vectors)
drop = keras.layers.Dropout(rate=0.2)(inner)
outputs = keras.layers.Dense(10, activation='softmax')(drop)
model = keras.Model(inputs, outputs)
```

```
learning_rate = 0.005
optimizer = keras.optimizers.Adam(learning_rate=learning_rate)
loss = keras.losses.CategoricalCrossentropy()
```

Next, we will compile the model with the optimizer and loss to achieve the best accuracy:

```
model.compile(optimizer=optimizer, loss=loss, metrics=["accuracy"])
```

Now, we will train the network with the clothing dataset:

```
model.fit(train_ds, validation_data=validation_ds, epochs=10)

Epoch 1/10
97/97 [==============================] - 18s 156ms/step - loss: 1.1373
- accuracy: 0.6228 - val_loss: 0.7507 - val_accuracy: 0.7830
...
Epoch 10/10
97/97 [==============================] - 12s 118ms/step - loss: 0.1951
- accuracy: 0.9289 - val_loss: 0.7641 - val_accuracy: 0.7918
```

Looking at these results, it is clear that the model is overfitted since the model achieved 92.7% accuracy on the training data and only 81.52% accuracy on the validation data, which is almost a 10-11% difference at the end of 10 epochs.

We could try training dense layers with different learning rates to obtain a more generalized model. A learning rate controls how quickly we want the model to learn from training data and adjust its weights to fit the data.

A low learning rate is like watching a video at a slow pace to ensure most of the details are covered, whereas a high learning rate is like watching a video at a faster pace and some details may be missed. An optimized learning rate is crucial for successful training, but it often requires tuning to find a balance between convergence speed and accuracy.

Another way to reduce overfitting is by adjusting the dropout rate, which is a regularization technique in which a percentage of data is omitted at random. Both adjustments require experimentation and are more model-centric approaches.

Following a data-centric approach, we want to test more data examples or better data examples. We need examples where the model doesn't try to memorize specific pixels so that if it sees 120 red in the 130th-pixel location, it starts believing the image is of pants. Hence, to ensure a well-generalized model, we could leverage data augmentation.

Concerning data-centric AI, data augmentation can be referred to as a technique to artificially increase the size and diversity of a training dataset by applying various transformations to the existing data.

These transformations can include rotation, scaling, cropping, flipping, adding noise, and many others, depending on the type of data being augmented.

There are some common augmentation techniques, such as creating different angles of the image, shifting images, zooming images in and out, flipping them upside down, and more. However, before applying augmenting techniques, we must first consider different ways data will be generated. For instance, if users don't generate pictures upside down, then augmenting images to create flipped images may only add noise and not provide a good signal to the model. For the following example, we will apply zoom augmentation, where images will be zoomed in and out a little, some shifting where clothing in the images will shift close to the edges, and apply vertical flips so that if some images are taken using a mirror, we can capture additional data for each scenario. We will achieve this by updating the image generator function, adding these extra parameters, and then training the model on the best parameters.

We will apply a rotation range of 10, a shear range of 10, a width and height shift range of 0.2, a zoom range of 0.1, and vertical flip. To achieve this, we will tweak the `ImageDataGenerator` function as this will create more examples of the training data under the hood:

```
train_gen = ImageDataGenerator(preprocessing_function=preprocess_
input,
                               rotation_range=10,
                               shear_range=10,
                               width_shift_range=0.2,
                               height_shift_range=0.2,
                               zoom_range=0.1,
                               vertical_flip=True)

train_ds = train_gen.flow_from_directory(directory='./clothing-
dataset-small/train/', target_size=(150,150), batch_size=32, seed=42)

val_gen = ImageDataGenerator(preprocessing_function=preprocess_input)
validation_ds = val_gen.flow_from_directory(directory="./clothing-
dataset-small/validation/", target_size=(150,150), batch_size=32,
shuffle=False)

Found 3081 images belonging to 10 classes.
Found 341 images belonging to 10 classes.
```

We will also create a function that will take in the learning rate and return the model and its parameters:

```
def make_fashion_classification_model(learning_rate: float=0.1):
    """Function to create a dense custom model"""
    #### Base model ####
    inputs = keras.Input(shape=(150,150,3))
    base = base_cnn_model(inputs)
```

```
    vectors = keras.layers.GlobalAveragePooling2D()(base)

    #### Dense model ####
    outputs = keras.layers.Dense(10, activation='softmax')(vectors)
    model = keras.Model(inputs, outputs)

    #### Optimizing the model ####
    optimizer = keras.optimizers.Adam(learning_rate=learning_rate)
    loss = keras.losses.CategoricalCrossentropy()
    model.compile(optimizer=optimizer, loss=loss,
metrics=["accuracy"])

    return model
```

Next, we will use this function and pass 0.005 as the learning rate and add 50 epochs since we have generated a lot more data for augmentation:

```
model = make_fashion_classification_model(learning_rate=0.005)
model_run = model.fit(train_ds, validation_data=validation_ds,
epochs=50)

Epoch 1/50
97/97 [==============================] - 28s 266ms/step - loss: 1.5281
- accuracy: 0.4920 - val_loss: 0.9664 - val_accuracy: 0.6686
...

Epoch 32/50
97/97 [==============================] - 25s 253ms/step - loss: 0.7445
- accuracy: 0.7491 - val_loss: 0.6161 - val_accuracy: 0.8123
```

The model achieved lower training accuracy, but the model is more generalizable and not overfitted. Also, note that the best validation accuracy was achieved at the 32nd epoch. However, it is difficult to note at which epoch a model will achieve the best accuracy.

To achieve this, we can further utilize the model checkpoint functionality to ensure only a model that achieves a minimum validation accuracy of 78% at a given epoch will be created and saved, and only when a previous best accuracy is surpassed will a new model be created. We can then use these saved models to score test data.

In the next step, we will add a dropout rate of 0.2 and an inner layer of 50. We will update the function for training the network and add the checkpoint functionality. Once the checkpoint has been created, we'll add it as a callback to the `fit` function.

We encourage you to utilize hyperparameterization with the learning rate, dropout rate, and inner layer while using the checkpoint to ensure the best accuracy.

First, we will define a function that takes three inputs – the learning rate, the inner layer, and the dropout rate:

```
def make_fashion_classification_model(learning_rate: float=0.001,
inner_layer: int=50, drop_rate: float=0.2):
    """Function to create a dense custom model with learning rate,
inner layer and dropout rate"""

    #### Base model ####
    inputs = keras.Input(shape=(150,150,3))
    base = base_cnn_model(inputs)
    vectors = keras.layers.GlobalAveragePooling2D()(base)

    #### Dense model layers ####
    inner = keras.layers.Dense(inner_layer, activation='relu')
(vectors)
    drop = keras.layers.Dropout(rate=drop_rate)(inner)
    outputs = keras.layers.Dense(10, activation='softmax')(drop)
    model = keras.Model(inputs, outputs)

    #### Optimizing the model ####

    optimizer = keras.optimizers.Adam(learning_rate=learning_rate)
    loss = keras.losses.CategoricalCrossentropy()
    model.compile(optimizer=optimizer, loss=loss,
metrics=["accuracy"])

    return model
```

Next, we will define the checkpoint. This is where we will save all the models with various epochs where the validation accuracy has reached a minimum of 78%. Then, we will add this checkpoint to the callback of the fitting function:

```
checkpoint_model = keras.callbacks.ModelCheckpoint(
    filepath="xception_v1_{epoch:02d}_{val_accuracy:.4f}.h5",
    monitor="val_accuracy",
    save_best_only=True,
    initial_value_threshold=0.8,
    mode="max")

model = make_fashion_classification_model()
model.fit(train_ds, validation_data=validation_ds, epochs=50,
callbacks=[checkpoint_model])
```

```
Epoch 1/50
97/97 [==============================] - 28s 266ms/step - loss: 1.4822
- accuracy: 0.5138 - val_loss: 0.8920 - val_accuracy: 0.7155
...

Epoch 37/50
97/97 [==============================] - 24s 250ms/step - loss: 0.5789
- accuracy: 0.7916 - val_loss: 0.5809 - val_accuracy: 0.8123
```

Now that the best model has been trained and saved, we will import the saved model, score all the data, and calculate the test accuracy:

```
model = keras.models.load_model('xception_v1_37_0.8123.h5')
test_gen = ImageDataGenerator(preprocessing_function=preprocess_input)
test_ds = test_gen.flow_from_directory(directory="./clothing-dataset-
small/test/", target_size=(150,150), batch_size=32, shuffle=False,
seed=42)

accuracy = model.evaluate(test_ds)[1]
print(f"accuracy on train data was 79.16, where as validation accuracy
was 81.23, but test accuracy is {accuracy*100:.2f}")
Found 372 images belonging to 10 classes.
12/12 [==============================] - 2s 94ms/step - loss: 0.6317 -
accuracy: 0.7715
accuracy on train data was 79.16, whereas validation accuracy was
81.23, but test accuracy is 77.15
```

The model that we used to predict the test data had a training accuracy of 79.16% and a validation accuracy of 81.23%. The test accuracy we achieved was only 77.15%, which can be improved iteratively – but beyond the scope of this example – by utilizing hyperparameter tuning for data augmentation parameters and model-centric parameters, which is encouraged in real life. However, due to computing and time constraints, this is outside the scope of this book.

Next, we will extract the target labels and build a function to provide a predicted probability score along with the relevant classes. First, we will load the image and preprocess it, and then we will score it:

```
labels = [i for i in train_ds.class_indices.keys()]

def preprocess_image(path: str, target_size: tuple):
    """Function to preprocess image"""

    img = load_img(path=path, target_size=target_size)
    img_input = np.array([np.array(img)])
    preprocessed_image = preprocess_input(img_input)
    return preprocessed_image
```

```
def decode_predictions(pred):
    """Function to decode prediction"""

    result = {c: format(float(p), '.8f') for c, p in zip(labels,
pred)}
    final_prediction = sorted(result.items(), key=lambda
x:x[1],   reverse=True)[0]
    return final_prediction
```

Next, we will extract a random image of pants and run it through the model to get the prediction:

```
path = './clothing-dataset-small/train/pants/188eaa2d-1a69-49b1-a9fb-
7b3789ac93b4.jpg'
load_img(path)
```

This will result in the following output:

Figure 7.11 – A randomly selected pair of pants from the clothing dataset

Next, we will load and preprocess the image before scoring it. Finally, we will decode the predictions to extract the final prediction. For this, we will leverage the functions we created earlier:

```
preprocessed_image = preprocess_image(path=path, target_
size=(150,150))
preds = model.predict(preprocessed_image)
```

```
results = decode_predictions(preds[0])
results


1/1 [==============================] - 0s 26ms/step
('pants', '0.99980551')
```

According to the model, the image has a probability of 99% to be classified as a pair of pants, which is quite accurate.

In this section, we have been able to demonstrate, through the data-centric technique of data augmentation, how to generalize the model. We believe that by iterating over data augmentation parameters, we can further improve the quality of the model. Once the parameters have been tuned, we recommend that practitioners combine model-centric techniques such as regularization, the learning rate, inner layers, and the dropout rate to further tune and improve the model.

We will now move on to exploring another topic that relies on unstructured data: synthetic data for text and natural language processing.

Natural language processing

Synthetic text data is typically used to increase the depth and breadth of written words and sentences with a similar semantic meaning to real-life observations.

The most common augmentation techniques that are used to create synthetic data for natural language processing involve replacing words with synonyms, randomly shuffling the position of words in a sentence, and inserting or deleting words in a sentence.

For example, the sentence "I love drinking tea" could be transformed into "I take great pleasure in consuming tea" without losing the contextual meaning of the statement. This is an example of *synonym replacement*, where "love" has been replaced with "take great pleasure in" and "drinking" has been replaced with "consuming."

Back translation is another NLP technique that involves translating a sentence in one language into another language, and then back into the original language. Often, this will generate slightly different sentence structures with a similar semantic meaning, which makes it a great way to combat overfitting while increasing the size of your training data.

We will illustrate a simple example of how to perform back translation using *Hugging Face Transformers* – specifically, the *MarianMT* suite of language models. MarianMT models were first created by Jörg Tiedemann using the Marian C++ library for fast training and translation but are now offered through the Hugging Face suite of Python libraries.

At the time of writing, the resource offers 1,440 transformer encoder-decoder models, each with six layers. These models support various language pairs, based on the *Helsinki-NLP* framework developed by the Language Technology Research Group at the University of Helsinki in Finland.

In this example, we want to translate the following three sentences from English into Spanish, and then use the same technique to translate the Spanish sentences back into English:

- The man glanced suspiciously at the door
- Peter thought he looked very cool
- Most individuals are rather nice

The goal is to generate similar sentences with the same semantic meaning but slightly different wording as this synthetic data will give our eventual model more data to learn from.

First, we'll install the required Python libraries:

```
pip install transformers sentencepiece
pip install mosestokenizer sacremoses
```

Then, we'll import the `MarianMTModel` and `MarianTokenizer` packages from the `transformers` library and define our input text string as `src_text`:

```
from transformers import MarianMTModel, MarianTokenizer
src_text = ['The man glanced suspiciously at the door', 'Peter thought
he looked very cool', 'Most individuals are rather nice']
```

Now, we'll define our `translator` model using the `Helsinki-NLP/opus-mt-en-es` language model, which translates from English into Spanish. The `MarianTokenizer` and `MarianMTModel` functions are used to define and execute our tokenizer and translation model, respectively.

The final output is stored as `trans_out`, which is then used as the input for our back translation model:

```
translator = 'Helsinki-NLP/opus-mt-en-es'
tokenizer = MarianTokenizer.from_pretrained(translator)

model = MarianMTModel.from_pretrained(translator)
translated = model.generate(**tokenizer(src_text, return_tensors="pt",
padding=True))
trans_out = [tokenizer.decode(t, skip_special_tokens=True) for t in
translated]
```

In this basic example, we simply repeat the same modeling exercise in reverse to produce slightly altered versions of the original input sentences. We use 'Helsinki-NLP/opus-mt-es-en' to translate back into English:

```
back_translator = 'Helsinki-NLP/opus-mt-es-en'
tokenizer = MarianTokenizer.from_pretrained(back_translator)

model_es = MarianMTModel.from_pretrained(back_translator)
back_translated = model_es.generate(**tokenizer(trans_out, return_
```

```
tensors="pt", padding=True))
[tokenizer.decode(t, skip_special_tokens=True) for t in back_
translated]
```

The following table shows the original input sentences against the model's output. The generated sentences have slightly different wording, but generally, they have the same semantic meaning as the originals. To use these sentences in a training dataset for a supervised model, they must inherit the labels of their original "parent" sentences:

Input Sentence	Back Translation
The man glanced suspiciously at the door	The man looked suspiciously at the door
Peter thought he looked very cool	Peter thought he looked great
Most individuals are rather nice	Most individuals are quite pleasant

Table 7.1 – Examples of back translation using Hugging Face Transformers

Privacy preservation

Synthetic data is also extremely useful for protecting the privacy and identity of individuals. The main aim of using synthetic data for privacy preservation is to make it impossible to identify individuals in a dataset while still keeping the statistical properties of the original dataset (close to) intact.

Synthetic data is an excellent option for privacy preservation since it allows information to be shared without revealing private or sensitive information. To achieve this, we must create data that resembles the original but does not contain any personally identifiable information.

The use of synthetic data allows organizations to share data for research or other purposes without compromising the privacy of individuals. There are several benefits to using synthetic data for privacy preservation – for example, you can reduce the risk of data breaches since the data contains no personal or sensitive information.

Data privacy regulations around the world are increasingly making it mandatory to protect individuals' privacy when using consumer data for analytical purposes. Synthetic data can be used to comply with privacy regulations, such as GDPR in the European Union or the HIPAA privacy rule in the United States, which sets standards for protecting personal data and preventing it from being shared without consent.

In general, synthetic data is a useful tool for preserving privacy because it allows organizations to share data without revealing sensitive or personal information. It is particularly useful for managing the trade-off between data quality and individual privacy in machine learning, making it an integral part of the data-centric toolbox.

Consider, for example, a bank that wants to use sensitive customer data for analytical activities such as churn modeling, fraud detection, and credit assessments. Using customer data for these activities typically brings about many compliance risks and mandated requirements that must be managed to avoid privacy breaches and heavy fines from regulators.

By having pre-generated synthetic datasets at hand, data scientists from various parts of the business can quickly and safely build models that would yield similar results to models built on real-world data. By using *appropriately constructed* synthetic data, the organization avoids going through cumbersome compliance and governance processes every time a new model is built and productionized.

However, this doesn't mean the use of privacy-preserving data is without its risks. It may happen, for example, that the generative model overfits the original data and produces synthetic instances too close to the original data.

Also, although synthetic data may appear anonymous, there may be instances where sophisticated hacks can reveal the identities of individuals. The aim of privacy-preserving synthetic data is to limit the risk of this happening.

Let's explore some common privacy disclosure scenarios to understand these risks and how we might limit them.

Types of privacy disclosure

To further understand and appreciate the usefulness of synthetic data for privacy preservation, let's have a look at three different types of privacy disclosure that can occur. This is not an exhaustive list of potential disclosure events, but it does help build an understanding of the potential and limitations of using synthetic data for this purpose.

Direct identity disclosure is the most obvious type of privacy disclosure. This is where an external adversary, such as a hacker, tries to gain information by matching the identity of an individual to records of private information. An example of this could be matching a person's identity with medical records.

Inferential identity disclosure is a form of data privacy breach where certain pieces of personal information can be derived from data that has been made publicly available, without explicitly revealing an individual's identity. This type of privacy breach occurs when an attacker uses statistical analysis to infer characteristics about an individual by analyzing patterns and correlations in a dataset. For example, an attacker may be able to determine the gender of an individual from publicly available data by analyzing patterns between particular characteristics and the corresponding gender.

Fully synthetic data, by design, makes direct identity disclosure almost impossible. However, an attacker could use the analysis of a synthetic dataset to infer information about a particular group of people, despite not being able to identify individuals in the dataset.

For example, say an original dataset contains sensitive medical information. The synthetic version of this data preserves the same statistical properties as the original data. With basic statistical methods or more advanced machine learning models, an adversary can identify groups of people with similar characteristics and deduce their risk of a certain disease. An adversary could then leverage this knowledge to infer the risk of any individual that shares those same characteristics, without any direct identity disclosure taking place.

Another example of inferential disclosure is when an attacker can infer someone's financial status or income level based on certain behaviors, such as shopping habits or credit card usage patterns. In addition, an attacker may be able to determine the medical history of a person by analyzing health insurance claims and other related records. This can lead to serious consequences, such as discrimination or exploitation of sensitive information. Therefore, organizations must take appropriate steps to protect their data from inferential disclosure to ensure that it is not used for malicious purposes.

Membership inference attacks are similar to inferential disclosure, but they are not exactly the same. Rather than inferring personal information about an individual based on a group that shares similar characteristics, membership inference attacks aim to deduce if an individual who was present in the original dataset was used to create the synthetic dataset. This presents a huge privacy risk as, for example, it may reveal that someone has a certain illness without ever having disclosed their medical information. Preventing these attacks through synthetic data is difficult as the statistical properties of the original dataset have been maintained.

In other words, synthetic data is a potent weapon against direct identity disclosure but does not remove the risk of identity disclosure entirely. Let's examine why synthetic data is still superior to traditional identity-masking techniques.

Why we need synthetic data for privacy preservation

Traditional data de-identification techniques rely on two main approaches:

- **Anonymization**: This is the simplest form of de-identification and is where columns containing direct (customer ID, name, address) and quasi-identifiers (ZIP code, birth date) and other sensitive information are removed, hashed, encrypted, or masked. Metrics such as k-anonymity, l-diversity, and t-closeness are then used to validate the level of privacy preservation in a given dataset.

- **Differential privacy**: An algorithm for differential privacy uses statistical distributions such as Gaussian and Laplace to add randomly generated noise to the identifying features in a dataset. As a consequence, individuals' privacy will be protected because identifying information is concealed behind the noise.

Although these techniques lower the risk of individuals being identified directly, they aren't necessarily enough to completely remove it.

A 2019 study by Rocher et al[11] demonstrated that 99.98% of Americans could be re-identified with no more than 15 demographic attributes based on a sample size of the population of Massachusetts. The authors conclude that "*heavily sampled anonymized datasets are unlikely to satisfy the modern standards for anonymization set forth by GDPR and seriously challenge the technical and legal adequacy of the de-identification release-and-forget model.*"

Another study, this time by Sweeney, 2000,[12] found that 87% of the population in the US had reported characteristics that likely made them unique based only on ZIP code, gender, and date of birth. 53% of the US population is identifiable by only location, gender, and date of birth, where "location" is the city, town, or municipality where the person lives. 18% of the population are identifiable based on a combination of their county, gender, and date of birth.

In other words, it is quite possible to identify unique individuals based on only a few quasi-identifiers. By using synthetically generated data, we can remove these individual combinations from the dataset while preserving the overall statistical properties of the data.

Let's examine how this is done.

Generating synthetic data for privacy preservation

When we create synthetic data for privacy preservation, we have three goals:

- To maintain the utility of the original data by reflecting its statistical properties in the synthetic dataset.

- To ensure the data structure is the same as the original data. This means that we can use the same code and tools on synthetic data as on the original data, without needing to change anything.

- It should not be possible to tell which real-world individuals were part of the original dataset when using privacy-preserving synthetic data.

It is worth noting that there are different ways to create synthetic data. Partial synthetic data just replaces some of the data with synthetic data, while fully synthetic information is created from scratch, without any of the original data.

Depending on the approach taken, fully synthetic information can provide a stronger guarantee against personal identity breaches, without sacrificing much in terms of usability and convenience.

A great way for you to start practicing synthetic data generation is through the tools created by the **Synthetic Data Vault** (**SDV**) project. The project was first established by MIT's Data to AI Lab in 2016 and is a comprehensive ecosystem of Python libraries that allows users to learn single-table, multi-table, and time series datasets, which can then be used as the basis for generating synthetic data that replicates the format and statistical properties of the original data.

Thanks to this project, it's possible to easily supplement, augment, and – in some cases – replace real data with synthetic data when training machine learning models. Additionally, it enables machine learning models or other data-dependent software systems to be tested without the risk of exposure that comes with sharing actual data.

The SDV suite is comprised of several probabilistic graphical modeling and deep learning-based techniques. They are used to generate hierarchical generative models and recursive sampling algorithms, which enable synthetic versions of a variety of data structures.

We will use two different techniques from the SDV suite – GaussianCopula and CopulaGAN – to illustrate how to generate synthetic data for privacy preservation purposes. Then, we'll briefly look at how to measure the *quality* and *score* using metrics and charts.

GaussianCopula

A copula is a tool that's used to measure the dependence among random variables. GaussianCopula is a collection of multiple (that is, multivariate) normally distributed pieces of data. Taken together as one set, the copula lets us describe how these independent normal distributions are related by showing how changes in one element in the set affect the others – that is, their *marginal distributions*. This is important because this exercise aims to augment any one unique combination of variables while preserving the overall statistical properties of the dataset.

Example GaussianCopula Python program

Now, we will write a sample program to show how this works. It will do the following:

1. Load a sample dataset and then calculate the GaussianCopula model.

2. Use that model to generate some sample data given the GaussianCopula model.

3. Visualize the output and its statistical properties to understand how our model is performing.

First, we must install the necessary Python packages – sdv and pandas:

```
pip install sdv
pip install pandas
```

For this exercise, we will use the publicly available *Adult* dataset, also known as the *Census Income* dataset. To get started, download it from https://archive.ics.uci.edu/ml/machine-learning-databases/adult/adult.data.

We'll use standard pandas functions to create a DataFrame, df, from the preceding URL:

```
names = ['age', 'workclass', 'fnlwgt', 'education', 'education-num',
                'marital-status', 'occupation', 'relationship',
'race', 'sex',
                'capital-gain', 'capital-loss', 'hours-per-week',
                'native-country', 'income']

url = 'https://archive.ics.uci.edu/ml/machine-learning-databases/
adult/adult.data'
df = pd.read_csv(url, header=None, names=names, na_values=['?', ' ?'])
```

This will output the following DataFrame:

	age	workclass	fnlwgt	education	education-num	marital-status	occupation	relationship	race	sex	capital-gain	capital-loss	hours-per-week	native-country	income
0	39	State-gov	77516	Bachelors	13	Never-married	Adm-clerical	Not-in-family	White	Male	2174	0	40	United-States	<=50K
1	50	Self-emp-not-inc	83311	Bachelors	13	Married-civ-spouse	Exec-managerial	Husband	White	Male	0	0	13	United-States	<=50K
2	38	Private	215646	HS-grad	9	Divorced	Handlers-cleaners	Not-in-family	White	Male	0	0	40	United-States	<=50K
3	53	Private	234721	11th	7	Married-civ-spouse	Handlers-cleaners	Husband	Black	Male	0	0	40	United-States	<=50K
4	28	Private	338409	Bachelors	13	Married-civ-spouse	Prof-specialty	Wife	Black	Female	0	0	40	Cuba	<=50K

Figure 7.12 – The first five rows of the Adult dataset – our input dataset for synthetic data generation

With our DataFrame created, we will import `SingleTableMetadata`, which is a class that provides methods to manage metadata about a single table of data, such as the names and types of columns, relationships between columns, and more. SDV's modeling suite needs this metadata object as input.

Then, we will use the `detect_from_dataframe()` method to analyze the pandas `df` DataFrame and automatically detect and set metadata about the table.

Finally, we will load the appropriate APIs and objects from SDV and instantiate the `GaussianCopula` model. Then, we will use the `fit()` method to generate the model:

```
from sdv.metadata import SingleTableMetadata
metadata = SingleTableMetadata()
metadata.detect_from_dataframe(df)

from sdv.single_table import GaussianCopulaSynthesizer

gc_model = GaussianCopulaSynthesizer(metadata)
gc_model.fit(df)
```

Normally, we would take a sample of the input dataset that is smaller than the full input dataset to generate the model, but in this case, we'll take the entire input data since it isn't too large.

Now, let's generate the synthetic dataset:

```
gc_synthetic = gc_model.sample(num_rows=df.shape[0] )
```

Once we have generated our synthetic data, we can assess its quality by comparing it to the attributes of the real data. This can be done by using several quality metrics.

Let's go ahead and do that.

Calculating quality scores

To measure the quality of the synthetic data, we can use various *score* metrics from the *SDV* package. The definition and interpretation of the scores vary depending on which metric we are looking at.

Let's look at some relevant metrics for measuring statistical similarity between original and synthetic data, as well as the risk of inference attacks being successful. Scores range between 0 and 1. The interpretation of 0 or 1 varies according to what metric you are using:

- BoundaryAdherence: This describes whether the synthetic data lies within the range of the max and min for a column in the real data. 1 means yes and 0 means no.

- StatisticSimilarity: This compares the mean, median, and standard deviation in a column between real and synthetic data.

- CategoricalCAP: This is the risk of disclosing private information using an inference attack – that is, a hacker knows some of the real data and can match it up with the synthetic. A score of 1 means there is a high risk.

- Data Likelihood: This calculates how likely it is that the data will match observations in the original data. This is similar to the Detection metric, which asks whether the machine learning model can tell which is the original dataset and which is the fabricated one.

- KSComplement: This shows whether the column shape of the real and synthetic data are the same using the **Kolmogorov-Smirnov (K-S)** test. The K-S test measures the maximum distance between the **cumulative distribution function (CDF)** of the two datasets. However, it uses its complement (the 1 - KS statistic).

- MissingValueSimilarity: This measures the proportion of missing data in the real and synthetic datasets.

The code for showing all of these metrics is nearly the same. Simply call the appropriate package, then run the compute method:

```
from sdmetrics.single_column import CategoryCoverage

CategoryCoverage.compute(
    real_data=df['workclass'],
    synthetic_data=synthetic['workclass']
)
```

Here is an example for `MissingValueSimilarity`:

```
from sdmetrics.single_column import MissingValueSimilarity

MissingValueSimilarity.compute(
    real_data=df['marital-status'],
    synthetic_data=synthetic['marital-status']
)
```

The output score is equal to 1.0, which means the model has successfully matched the proportion of missing values in the synthetic dataset.

Quantifying and visualizing data quality

We also want to quantify and visualize the quality of our synthetic data compared to the original set. For this purpose, we'll use the *diagnostic* and *quality* reports from the SDV library.

The diagnostic report should always produce a score of 100%, which tells us that primary keys are unique and non-null, continuous values in the synthetic data adhere to the min/max range in the original data, discrete values line up with the same categories across real and synthetic data, and column names are the same:

```
from sdv.evaluation.single_table import run_diagnostic

diagnostic = run_diagnostic(
    real_data=df,
    synthetic_data=gc_synthetic,
    metadata=metadata
)
```

Here's our output:

```
Overall Score: 100.0%

Properties:
- Data Validity: 100.0%
- Data Structure: 100.0%
```

The SDV quality report evaluates how well your synthetic data captures the mathematical properties of our original real data. It does this through a set of metrics that measure various aspects of the *fidelity* between the two datasets. **Data fidelity** refers to how accurate a dataset is at representing the features of its source.

The report provides an overview of the results, as well as detailed visualizations and explanations for each metric so that you can quickly understand the strengths and weaknesses of your synthetic data. By understanding how well your synthetic data captures the mathematical properties of the real data, you can take steps to improve it if needed.

The SDMetrics quality report is a valuable tool that helps you ensure your synthetic data is as accurate and reliable as possible. Here's how we can use it:

```
from sdv.evaluation.single_table import evaluate_quality
quality_report = evaluate_quality(
    real_data=df,
    synthetic_data=gc_synthetic,
    metadata=metadata
)
```

The preceding code produces a quality report with various metrics and visualizations that show the overall similarities between the original and synthetic data:

```
Overall Score: 84.6%

Properties:
- Column Shapes: 87.57%
- Column Pair Trends: 81.63%
```

Here are a couple of important metrics to know about:

- Column Shapes: A column's shape tells us how data is distributed. A higher score means that the real and synthetic data are more similar. A separate column shape score for every column is calculated, but the final score is the average of all columns.

- Column Pair Trends: The correlation between two columns indicates how their trends compare to each other; the higher the score, the more similar those trends are. A score is produced for each column pair in the data, while the final score is the average of all columns. This is an important score that tells us whether our synthetic data has captured the relationships between variables in the original dataset.

We can also visualize the dimensions of these metrics with the get_visualization command:

```
fig = quality_report.get_visualization(property_name='Column Pair Trends')
fig.show()
```

This will generate the following plot:

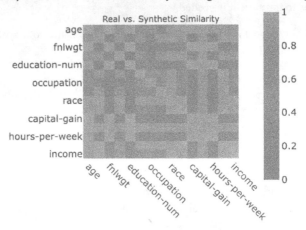

Figure 7.13 – A correlation matrix comparing column pair trends
between the original and synthetic datasets

As you can see, most column pairs have a high similarity score, but the `capital-gain` column is far apart. We can use the following code to visualize the real and synthetic `capital-gain` column distributions side by side:

```
from sdv.evaluation.single_table import get_column_plot

fig = get_column_plot(
    real_data=df,
    synthetic_data=gc_synthetic,
    metadata=metadata,
    column_name='capital-gain'
)

fig.show()
```

The output is generated as follows:

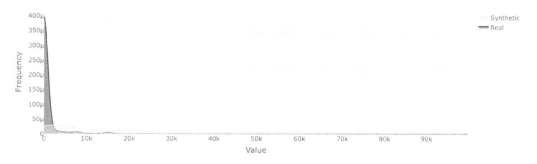

Figure 7.14 – A comparison of the distribution of the real and synthetic capital-gain columns. The GaussianCopula model hasn't done a good job of matching the distribution

In this case, we would test various column distribution functions to find a better match for this particular column. In this example, we used the `GaussianCopula` function to create a synthetic dataset. However, the SDV library contains several other distributions that can be useful, depending on the characteristics of your original dataset. Let's explore how to change the default distribution.

Varying column distribution functions

The `GaussianCopula` function determines which statistical distribution best describes each copula, but it doesn't always get it right. Luckily, we can override the preselection and pick our preferred distribution.

We have the following choices:

- **Gaussian (normal) distribution**: Use this if your data is continuous and symmetrically distributed around the mean. It's often used for naturally occurring data, such as the heights or weights of a population.

- **Gamma distribution**: This is used for positive-only, skewed data. It's often used for things such as wait times or service times.

- **Beta distribution**: This is used for variables that are bounded between 0 and 1, such as proportions or probabilities.

- **Student's t-distribution**: This is similar to the Gaussian distribution but has heavier tails. It's often used when the sample size is small or the standard deviation is unknown.

- **Gaussian kernel density estimation (KDE)**: Use this for non-parametric data – that is, when you don't know or want to assume a specific distribution. The KDE uses the data itself to estimate its distribution.

- **Truncated Gaussian distribution**: Use this when you have data that follows a Gaussian distribution but is bounded within a specific range.

First, here is how to show the distributions it calculated:

```
gc_model.get_learned_distributions()
```

This statement produces a detailed list of all the columns in the dataset. Instead of showing the output here, the model defaulted to a beta distribution for all columns.

To change a distribution function for a given column, just create a model again but this time explicitly apply a specific distribution to that column. In this case, we will apply a gamma distribution to the `capital-gain` column:

```
gc_model2 = GaussianCopulaSynthesizer(
    metadata,
    numerical_distributions={
        'capital-gain': 'gamma',
    })
gc_model2.fit(df)
```

The resulting output is a new synthetic dataset with a `capital-gain` column distribution much closer to the real data:

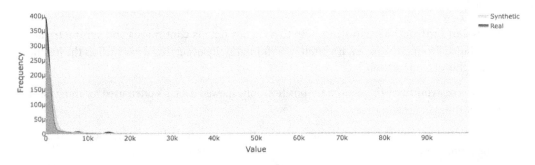

Figure 7.15 – A new comparison of the capital-gain columns. Using the gamma distribution on this column improved the similarity between the synthetic and original data

Another useful package from the SDV library is CopulaGAN. This algorithm is a blend of the GaussianCopula and CTGAN algorithms. Let's compare the performance of CopulaGAN to GaussianCopula on the *Adult* dataset.

CopulaGAN code example

CopulaGAN is a variation of GaussianCopula that can yield better results using a simplified GAN model. We will compare the two models in this section, but first, here is the code to generate CopulaGAN using the same input dataset and metadata object:

```
from sdv.single_table import CopulaGANSynthesizer

cg_model = CopulaGANSynthesizer(metadata)
cg_model.fit(df)
```

Measuring data quality from CopulaGAN

Now, let's look at data quality regarding the CopulaGAN model, repeating some of the same techniques we used with GaussianCopula:

```
quality_report = evaluate_quality(
    real_data=df,
    synthetic_data=cg_synthetic,
    metadata=metadata
)

Overall Score: 87.39%

Properties:
- Column Shapes: 91.75%
- Column Pair Trends: 83.04%
```

Figure 7.16 provides a visual representation of the column pair trends of the original and synthetic datasets:

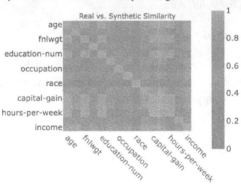

Figure 7.16 – Column pair trends for the Adult dataset using CopulaGAN

Understanding the difference between GaussianCopula and CopulaGAN

CopulaGAN is a hybrid AI model that combines the human accessibility of Gaussian copulas with the robust accuracy of GANs[13].

A GAN is a deep learning algorithm. If you've worked with neural networks, you know that they are a kind of black box, meaning the coefficients of the nodes in the network are functions and not numbers. They are very hard to explain or understand compared to, for example, a polynomial or linear model.

GaussianCopula is easier to explain. It works by trying different known statistical distributions (normal, Weibull, and others), which is very useful for known or easily observable distributions. Then, for each column, it picks the one that matches the closest.

The team behind the SDV project developed CopulaGAN to get the best of both worlds: a more accurate model that is still explainable.

The following table compares the results from our two models in the previous examples. CopulaGAN achieved a higher overall quality score because it was able to match the column shapes more:

	GaussianCopula	**CopulaGAN**
Overall Quality Score	84.6%	87.39%
Column Shapes	87.57%	91.75%
Column Pair Trends	81.63%	83.04%

Table 7.2 – Data quality comparison of the GaussianCopula and
CopulaGAN algorithms on the Adult dataset

Of course, quality is a highly complex topic and our example is not exhaustive in that regard. You would have to look at the scores across all columns and all the different types of scores to validate the accuracy of the synthetically generated data. In other words, you cannot say for definite that `CopulaGAN` is more accurate in all cases without doing a deeper review of the variables in the dataset. This is particularly important when you are dealing with high-stakes datasets and use cases.

One additional metric to consider is run speed. Anecdotally, when we wrote this example, `CopulaGAN` took 1 hour to complete, while `GaussianCopula` took 15 seconds to complete.

Validating the privacy of our new dataset

Now that we have constructed a synthetic dataset for our use case, we need to ensure we have prevented the ability to re-identify individuals from the original dataset.

To know the likelihood that individuals can be re-identified, we need an accurate measure of the difference or "distance" between the original and synthetic records. The farther the two are apart, the less probable that they can be identified as one entity. If we are discussing personal information in tabular form, we need a methodology for measuring the distance between qualitative and quantitative attributes alike.

To accurately measure the closeness of two rows within a dataset containing both qualitative and quantitative information, we can utilize a similarity coefficient called Gower's distance.

Gower's distance is a unique type of distance measure that differs from distance measures. It stands out in terms of its ability to calculate the difference between two entities with both numerical and categorical values. This is important because many common clustering algorithms, such as K-means clustering, only work when all of the variables are numeric.

Gower's distance returns a similarity coefficient between 0 (indicating identical observations) and 1 (showing that they are at the maximum distance).

Suppose we have a set of p features in the original (o) and synthetic (s) datasets:

- For ordinal numbers, the distance from one feature to the other is simply the absolute value of their difference divided by the range of that variable. We divide by the range to normalize the data so that large numbers won't be given greater weight than small ones.

- Categorical variables are turned into numbers so that we can do math. The formula is simple – if those values are the same, their distance is 0; otherwise, it is 1.

Gower's distance is the sum of the distances divided by the number of features – the average of the terms. Since we divide these differences by the number of features, this is the same as saying Gower's distance is the average distance.

Then, we make a definition for **closeness** and call it the **distance to the closest record**. For every element in *s*, the closest row in *o* is the one with the minimum Gower's distance. A distance of 0 means that two rows of data are the same, while a distance of 1 means that two rows are as different as possible given the observations in the dataset we're using.

Let's practice applying Gower's distance using the `Gower` Python package.

Gower's distance Python example

In our Gower's distance practice example, we'll use the *Adult* dataset and compare the synthetic output generated by `CopulaGAN` to the original data. We recommend using a small subset of the *Adult* dataset (for example, 1,000 rows) to practice as Gower's matrix calculation can take a long time to run on larger sets.

First, we'll create the DataFrame for our model based on the top 1,000 rows from the existing `df` DataFrame, which contains the full *Adult* dataset. Then, we'll fit a `GaussianCopula` model on this dataset and generate a new synthetic dataset called `synthetic`:

```
new_df = df.head(1000)

model = GaussianCopulaSynthesizer(metadata)
model.fit(new_df)

synthetic = model.sample(num_rows=df.shape[0] )
```

Next, we'll install the `Gower` package and calculate the Gower's distance matrix between our two datasets:

```
pip install gower

import gower

gowerMatrix=gower.gower_matrix(new_df, synthetic)
print(gowerMatrix)
```

This will generate results similar to the following, where the distance between each row in the dataset is calculated:

```
[[0.4460268  0.41092068 0.39990428 ... 0.30849126 0.33108675 0.5243855 ]
 [0.5248843  0.16020963 0.36551976 ... 0.4707944  0.36200836 0.46959284]
 [0.30875853 0.4095393  0.3763651  ... 0.36865947 0.30754757 0.367513  ]
 ...
 [0.3892014  0.43856084 0.4261839  ... 0.46071026 0.35872692 0.5520257 ]
 [0.37288257 0.26404738 0.31969768 ... 0.3632057  0.25088012 0.37751225]
 [0.4604928  0.2709656  0.39038438 ... 0.5448743  0.41212356 0.53527355]]
```

Figure 7.17 – The resulting gowerMatrix

Now, we'll use the `gower_topn()` function to find the top *n* (in this case, 10) closest (that is, most similar) rows.

To ensure the synthetic dataset passes our test, we must make sure none of its values are equal to 0; otherwise, this would indicate that some rows in the synthetic data resemble those from the original. Generally speaking, we want the top values to be sufficiently distanced from 0 as this reduces the risk of reidentification:

```
gower.gower_topn(df.iloc[:,], synthetic.iloc[:,], n = 10)
```

The result is the index of the top 10 closest rows and their Gower's distance. In this case, the smallest distance between two rows in our datasets is 0.02205, which means our synthetic dataset is not sufficiently different from the original at the individual row level:

```
{'index': array([26320, 29200, 18735, 24149, 18316, 22925,  4836,
15360, 42, 3523]),'values': array([0.02205753, 0.02578343, 0.03649067,
0.0374441 , 0.03785798, 0.04503146, 0.06345809, 0.08126822,
0.08237292, 0.08368524], dtype=float32)}
```

In this case, we would have more work to do to reduce the similarity between sets. Here are a few techniques you could use to achieve this:

- **Change the synthetic data generation process**: The first step would be to test different settings for `GaussianCopula` or test other synthesizers in the SDV catalog, such as CTGAN or `CopulaGAN`.

- **Add noise**: You can add random noise to the synthetic data. This will make the synthetic data more "unique" compared to the original data.

- **Perform feature transformation**: Apply some kind of transformation (for example, logarithmic, square root, exponential, and so on) to the features in the synthetic dataset.

- **Perform data augmentation**: Generate new synthetic data points that are not direct copies of the real data. You can do this by using techniques such as the **Synthetic Minority Over-Sampling Technique (SMOTE)** or **Adaptive Synthetic Sampling (ADASYN)**, both of which we'll discuss later in this chapter.

Remember that while your goal is to reduce similarity, you also want the synthetic data to be useful and representative of the real data. If you make the synthetic data too dissimilar, it may not serve its intended purpose.

As a side note, Gower's distance can also be used to find rows of data that are very similar to each other, which is useful for tasks such as creating lookalike audiences, clustering, or identifying at-risk populations.

For instance, imagine that you have just run a very successful email marketing campaign to a group of customers and you want to expand the campaign to customers who look like the ones in the original target group.

To do this, simply calculate Gower's distance between customers in the original target group and the rest of your customer base, and pick a target group based on the lowest Gower's distances.

Using synthetic data to improve model performance

In *Chapter 5, Techniques for Data Cleaning* and *Chapter 6, Techniques for Programmatic Labeling in Machine Learning*, we dealt with improving model performance by refining data quality. However, there are times when improving data quality may not be enough, especially when datasets are small. In such situations, we can take advantage of generating synthetic data to boost model performance.

As we covered previously in this chapter, synthetic data can help with generating more training examples, as well as generalizing the performance of the model by providing more examples of different variations and distributions of the data. Both of these uses can make the model more robust and less likely to overfit the training data.

With imbalanced datasets, a model gets biased toward the majority class as there are more examples of one class over another. This is the problem with the loan prediction dataset, where 30% of the data belongs to the minority class.

In this section, we will cover generating synthetic data for the minority class so that the model can generalize further and model performance metrics can improve. We will stick with a decision tree model and use synthetic data generation to further improve the signal strength of the data. We will use the `imblearn` library from Python to generate synthetic data.

Let's import the library and two oversampling methods, `SMOTE` and `ADASYN`, to oversample the minority class. We will also leverage the `Counter` method to count data samples pre- and post-synthetic data generation:

```
import imblearn
from imblearn.over_sampling import SMOTE, ADASYN
from collections import Counter
print(imblearn.__version__)
0.10.1
```

Both the SMOTE and ADASYN algorithms are used to generate synthetic data. However, ADASYN is more robust as it considers the density of points to generate synthetic data. SMOTE may generate synthetic data around the minority class, but it does so uniformly without considering how rare a data point is.

SMOTE creates synthetic samples by randomly selecting pairs of minority-class samples and interpolating new samples around the existing samples. This technique spreads out further into the space to increase the number of minority class samples. However, as the samples are chosen randomly, no weighting is given to rare sample points.

On the other hand, ADASYN considers rare data points in the feature space by computing the density distribution of the minority class samples. It generates synthetic samples in regions of the feature space where the density is low to ensure that synthetic samples are generated where they are most needed to balance the dataset. ADASYN uses the k-nearest neighbors algorithm to estimate the density distribution of the minority class samples. For each minority class sample, ADASYN computes the density based on the number of k-nearest neighbors that belong to the minority class. The value of k is a user-defined parameter, typically set to a small value such as 5 to 10. The density is the average distance from k nearest points. A higher average distance means lower density, and vice versa.

We iterate over different thresholds of the algorithm parameters, such as the number of nearest neighbors and the percentage of data to oversample. This helps us find the best parameters to generate the optimal number of samples so that both the test ROC and test accuracy get the maximum boost. Then, we combine those results in a DataFrame and choose the best parameters. This is done to measure the performance of the model that is using synthetic data generation.

For our example, we will only use ADASYN, but we encourage you to try different techniques, including SMOTE, for the problem at hand:

```
results = []
over_sampling = [0.65,0.7, 0.75, 0.8, 'auto']
n_neighbours = [1,3,5,7,9,10]
for os in over_sampling:
    for k in n_neighbours:
        oversample = ADASYN(random_state=1, sampling_strategy=os, n_
neighbors=k)
        counter = Counter(y_train)
        print(f"data size before applying smote technique is
{counter}")
        X_train_synthetic, y_train_synthetic = oversample.fit_
resample(X_train_transformed, y_train)

        counter = Counter(y_train_synthetic)
        print(f"data size after applying smote technique is
{counter}")

        model, test_predictions, train_roc, test_roc, train_acc, test_
acc = train_custom_classifier(
        X_train=X_train_synthetic,
        y_train=y_train_synthetic,
        X_test=X_test_transformed,
        y_test=y_test,
        clf=d_clf,
        params=d_param_grid)
```

```
        results.append((os, k, train_roc, test_roc, train_acc, test_
acc))


synthetic_df = pd.DataFrame(columns=['os_strategy', "n_neighbours",
"train_roc", "test_roc", "train_acc", "test_acc"], data=results)
```

Next, we must generate a DataFrame that contains our generated combinations of model parameters and model performance metrics and sort it by test accuracy. The DataFrame indicates that an oversampling strategy with a ratio of 75% for the minority to majority class, and a nearest neighbors value of 7, will provide the best accuracy and ROC score:

```
synthetic_df.sort_values(by="test_acc", ascending=False)
    os_strategy n_neighbours  train_roc   test_roc   train_acc   test_acc
15       0.75             7   0.881653   0.862913   0.837143   0.838710
21       0.8              7   0.881653   0.862913   0.837143   0.838710
7        0.7              3   0.854092   0.800490   0.813916   0.838710
1        0.65             3   0.854092   0.800490   0.813916   0.838710
25       auto             3   0.943918   0.718482   0.879288   0.822581
20       0.8              5   0.940859   0.777234   0.873381   0.806452
14       0.75             5   0.940859   0.777234   0.873381   0.806452
13       0.75             3   0.974384   0.763158   0.904011   0.790323
19       0.8              3   0.974384   0.763158   0.904011   0.790323
11       0.7             10   0.927560   0.766218   0.853073   0.790323
24       auto             1   0.938804   0.695226   0.868550   0.774194
28       auto             9   0.931854   0.750918   0.878415   0.758065
26       auto             5   0.935847   0.746634   0.880857   0.758065
29       auto            10   0.929612   0.747246   0.877660   0.758065
10       0.7              9   0.932205   0.752754   0.845070   0.758065
8        0.7              5   0.946023   0.754590   0.871914   0.741935
23       0.8             10   0.940005   0.738066   0.862166   0.741935
5        0.65            10   0.893740   0.805386   0.813472   0.741935
12       0.75             1   0.921313   0.744186   0.855051   0.725806
17       0.75            10   0.940240   0.716034   0.851641   0.725806
18       0.8              1   0.921313   0.744186   0.855051   0.725806
9        0.7              7   0.925762   0.767442   0.846039   0.725806
6        0.7              1   0.921313   0.744186   0.855051   0.725806
4        0.65             9   0.926310   0.774786   0.835314   0.725806
0        0.65             1   0.921313   0.744186   0.855051   0.725806
3        0.65             7   0.904908   0.786414   0.821718   0.709677
27       auto             7   0.955496   0.710526   0.879076   0.709677
2        0.65             5   0.890376   0.806610   0.810017   0.709677
16       0.75             9   0.956721   0.749694   0.882768   0.693548
22       0.8              9   0.956721   0.749694   0.882768   0.693548
```

Figure 7.18 – Output DataFrame

Now, we must apply the parameters from our highest-performing oversampling strategy and retrain the decision tree model:

```
counter = Counter(y_train)
print(f"data size before applying smote technique {tech_name} is
{counter}")
```

```
# transform the dataset
oversample = ADASYN(random_state=1, n_neighbors=7, sampling_
strategy=0.75)
X_train_synthetic, y_train_synthetic = oversample.fit_resample(X_
train_transformed, y_train)

counter = Counter(y_train_synthetic)
print(f"data size after applying smote technique {tech_name} is
{counter}")

model, test_predictions, train_roc, test_roc, train_acc, test_acc =
train_custom_classifier(
X_train=X_train_synthetic,
y_train=y_train_synthetic,
X_test=X_test_transformed,
y_test=y_test,
clf=d_clf,
params=d_param_grid)
data size before applying smote technique adasyn is Counter({1: 379,
0: 173})
data size after applying smote technique adasyn is Counter({1: 379, 0:
321})
Decision tree optimised
Getting the best params which are {'class_weight': 'balanced',
'criterion': 'entropy', 'max_depth': 6, 'max_features': 'sqrt', 'min_
samples_leaf': 3, 'min_samples_split': 30, 'random_state': 1}
Training roc is 0.8816528164788466, and testing roc is
0.8629130966952264
            training accuracy is 0.8371428571428572, testing_acc as
0.8387096774193549
```

ADASYN increased the number of minority class samples from 173 to 321 using synthetic data generation, which boosted the test accuracy to 83.8%. This is an almost 2% increase in accuracy. The ROC score was also boosted to 86.2%, which is a further increase of 4.4%.

These results demonstrate that synthetic data generation can provide significant gains in model performance, even for small datasets. However, it is important to note that this may not always be the case, especially if error analysis suggests that adding new data doesn't contribute to an improvement in model performance. In such cases, you may turn to collecting more data or features, or even performing feature engineering, before moving on to synthetic data generation.

When should you use synthetic data?

So far, we've established that synthetic data can be used for several purposes, but how do you decide whether to use synthetic data for your project or not?

For businesses seeking to gain an edge over their competitors through innovative or unconventional approaches, synthetic data provides an accessible middle ground between experimentation and reality. For governmental organizations wanting to learn from their vast stores of population data, synthetic data allows highly sensitive datasets to be analyzed without compromising individual privacy.

Experimentation and exploring the boundaries of your data (synthetic or real) can be incredibly valuable, but the benefit of introducing synthetic data should always be assessed against the cost and risk of making damaging predictions with that same data.

The central question is, "*What is the acceptable cost of an experiment?*," especially if it includes human collateral damage or reputational or financial loss.

In our opinion, synthetic data should be used when obtaining real-world data may be difficult, expensive, or unethical. The most common and practical use cases for synthetic data are for preserving the privacy of individuals and for creating simulations that are very difficult or impossible in traditional test environments. For these use cases, the benefits are more likely to outweigh the risks of using synthetic data, but that is not a guarantee, so make sure you manage risks appropriately.

The main risks to mitigate are perpetuation and exacerbation of bias. Machine learning models are inherently prone to overfitting and finding the "easiest" path through the data, so synthetic datasets should be rigorously tested to ensure they are fit for purpose.

Synthetic data can also accelerate the process of testing and training machine learning models, saving companies time and money in their development and deployment cycles. Furthermore, synthetic data is a useful tool for creating simulations that are not possible in traditional test environments.

Bear in mind that using synthetic data is typically just one of many avenues to take when building models or improving the accuracy of your predictions. It should only be used when the potential risk and effect on those impacted is understood and managed appropriately. On the other hand, if you can mitigate this risk – or in some cases, avoid "real-world" risks altogether – then it is a wonderful tool to have in your toolkit.

Summary

In this chapter, we provided a primer on synthetic data and its common uses. Synthetic data is a key part of the data-centric toolkit because it gives us yet another avenue to much better input data, especially when collecting new data is not feasible.

By now, you should have a clear understanding of the fundamentals of synthetic data and its potential applications. Synthetic data is often used for computer vision, natural language processing, and privacy protection applications. However, the potential of synthetic data goes well beyond these three realms.

Whole books have been dedicated to the topic of synthetic data and we recommend that you dive deeper into the subject if you want to become a true expert in synthetic data generation.

In the next chapter, we'll explore another powerful technique for improving your data without the need for collecting new data: programmatic labeling.

References

1. `https://datagen.tech/guides/synthetic-data/synthetic-data`, viewed on 12 November 2022

2. `https://blogs.gartner.com/andrew_white/2021/07/24/by-2024-60-of-the-data-used-for-the-development-of-ai-and-analytics-projects-will-be-synthetically-generated/`

3. `https://unity.com/our-company`, viewed on 15 November 2022

4. `https://venturebeat.com/ai/unitys-danny-lange-explains-why-synthetic-data-is-better-than-the-real-thing-at-transform-2021-2/`, viewed on 15 November 2022

5. Alcorn, M A et al 2019, *Strike (with) a Pose: Neural Networks Are Easily Fooled by Strange Poses of Familiar Objects*, viewed 13 November 2022: `https://arxiv.org/pdf/1811.11553.pdf`

6. `https://www.tesla.com/VehicleSafetyReport`, viewed 13 November 2022

7. Karras T, Aila T, Laine S, Lethtinen J, 2017, *Progressive Growing of GANs for Improved Quality, Stability, and Variation*: `https://arxiv.org/abs/1710.10196`

8. Karras T, Aila T, Laine S 2018, *A Style-Based Generator Architecture for Generative Adversarial Networks*: `https://arxiv.org/pdf/1812.04948.pdf`

9. Metz L, Poole B, Pfau D, Sohl-Dickstein J 2017, *Unrolled Generative Adversarial Networks*, ICLR 2017: `https://arxiv.org/pdf/1611.02163.pdf`

10. Jain N, Olmo A, Sengupta S, Manikonda L, Kambhampati S, 2021, *Imperfect ImaGANation: Implications of GANs Exacerbating Biases on Facial Data*, ICLR 2021 Workshop on Synthetic Data Generation – Quality, Privacy, Bias: `https://arxiv.org/pdf/2001.09528.pdf`

11. Rocher L, Hendrickx J M, de Montjoye Y A 2019, *Estimating the success of re-identifications in incomplete datasets using generative models*: `https://www.nature.com/articles/s41467-019-10933-3`

12. L. Sweeney, *Simple Demographics Often Identify People Uniquely, Carnegie Mellon University*, Data Privacy Working Paper 3. Pittsburgh 2000: `https://dataprivacylab.org/projects/identifiability/paper1.pdf`

13. `https://mobile.twitter.com/sdv_dev/status/1519747462088507393`, viewed on 25 January 2023

Techniques for Identifying and Removing Bias

In the realm of data-centric machine learning, the pursuit of unbiased and fair models is paramount. The consequences of biased algorithms can range from poor performance to ethically questionable decisions. It is important to recognize that bias can manifest at two key stages of the machine learning pipeline: data and model. While model-centric approaches have garnered significant attention in recent years, this chapter sheds light on the equally crucial data-centric strategies that are often overlooked.

In this chapter, we will explore the intricacies of bias in machine learning, emphasizing why data-centricity is a fundamental aspect of bias mitigation. We will explore real-world examples from finance, human resources, and healthcare, where the failure to address bias has had or could have far-reaching implications.

In this chapter, we'll cover the following topics:

- The bias conundrum
- Types of bias
- The data-centric imperative
- Case study

The bias conundrum

Bias in machine learning is not a novel concern. It is deeply rooted in the data we collect and the algorithms we design. Bias can arise from historical disparities, societal prejudices, and even the human decisions made during data collection and annotation. Ignoring bias, or addressing it solely through model-centric techniques, can lead to detrimental outcomes.

Consider the following scenarios, which illustrate the multifaceted nature of bias:

- **Bias in finance**: In the financial sector, machine learning models play a pivotal role in credit scoring, fraud detection, and investment recommendations. However, if historical lending practices favor certain demographic groups over others, these biases can seep into the data used to train models. As a result, marginalized communities may face unfair lending practices, perpetuating socioeconomic inequalities.

- **Bias in human resources**: The use of AI in human resources has gained momentum for recruitment, employee performance assessment, and even salary negotiations. If job postings or historical hiring data are biased toward specific genders, ethnicities, or backgrounds, the AI systems can inadvertently perpetuate discrimination, leading to a lack of diversity and inclusion in the workplace.

- **Bias in healthcare**: In healthcare, diagnostic algorithms are relied upon for disease detection and treatment recommendations. If training data predominantly represents certain demographics, individuals from underrepresented groups may receive suboptimal care or face delayed diagnoses. The implications can be life-altering, underscoring the need for equitable healthcare AI.

Now that we have covered areas where bias can arise, in the next section, we will cover the types of bias prevalent in machine learning.

Types of bias

In machine learning, there are generally five categories of bias that warrant attention. Although the list provided isn't exhaustive, these categories represent the most prevalent types of bias, each of which can be further subdivided.

Easy to identify bias

Some types of bias can be easy to identify using active monitoring and by conducting analysis. These include the following.

Reporting bias

This type of bias occurs when the data producers, data annotators, or data capturers miss out on important elements, which results in data not being representative of the real world. For instance, a healthcare business might be interested in patients' sentiments toward a health program; however, the data annotators may decide to focus on negative and positive sentiments, and sentiments that were neutral may be underrepresented. A model trained on such data will be good at identifying positive and negative sentiments but may fail to accurately predict neutral sentiments. This type of bias can be identified with active monitoring, where predictions on live data show drift from predictions on training data. To reduce reporting bias, it is important to articulate data points needed for the problem at the beginning phase of ML system design. It is also important to ensure that data used for training represents real-life data.

Automation bias

This type of bias occurs due to relying on automated ways of data collection and assuming data capture is not error-prone. As AI is becoming better, reliance on humans has significantly reduced, and hence it is often assumed that if an automated system is put in place, then it will magically solve all problems. Using active monitoring can help identify this type of bias, where the model accuracy is highly poor on real-life data. Another way of identifying this is by using humans to annotate labels and measure human performance versus algorithmic performance. As covered in *Chapter 6, Techniques for Programmatic Labeling in Machine Learning* systems fail and can lead to missing data or inaccurate data. AI is as good as the data it was trained on. One of the key principles of data-centricity is to keep humans in the loop; hence, when building automated systems, we should ensure the data generated represents real-world scenarios and data is diverse rather than overrepresented or underrepresented.

Selection bias

This type of bias occurs when data selected for training the model is not representative of real-life data. This bias can take multiple forms:

- **Coverage bias**: This bias can occur when data is not collected in a representative manner. This can happen when the business and practitioners are focused on outcomes, and ignore data points that do not contribute to the outcome. In healthcare, insurance companies may want to predict hospital admissions; however, data on people churning on insurance companies and using competitive insurance products, or data on people not claiming benefits to go into the hospital may not be readily available and, as a result, these groups of people may not be represented well in the training data.

- **Participation bias**: This bias can occur due to participants opting out of data collection processes, leading to one group being overrepresented over another group. For example, a model is trained to predict churn using survey data, where 80% of people who have moved to a new competitor are unlikely to respond to the survey, and their data is highly underrepresented in the sample.

- **Sampling bias**: This bias can occur when data collectors do not use proper randomization methods in data collection processes. For example, a model is trained to predict health scores based on survey data; instead of targeting the population at random, the surveyors chose 80% of people who are highly engaged with their health and are more likely to respond, compared to the rest of the responders. In the health industry, people who are more engaged with their health are likely to have a better health score than people who are less engaged, thus leading to a biased model toward healthy people.

Selection biases are difficult to identify; however, if drift is noted frequently in the data and highly frequent retraining is done to ensure the model quality does not degrade, then it will be a good time to investigate and check whether the data captured represents real-life data. Two types of analysis in regression modeling can help to identify this bias. One is conducting bivariate analysis, where a sensitive variable can be represented on the x axis and the target variable can be put on the y axis. If there is a strong association between the two variables, then it is important to evaluate the difference in the association metric at training time and post-scoring time. If the difference is significant, it is quite possible that the data used for training is not representative of real life. The second technique is to use multivariate analysis by comparing the possible outcomes when data is not fully represented and when data is fully represented. This can be done by separating the subgroups into data points that were included and the ones that were excluded at training time. We can run a multi-regression model by creating an independent variable group by labeling group 1 for data included and group 2 for data not included. We will add this new variable as a feature to the model training and then compare whether there is a significant difference in outcome between groups 1 and 2. If there is a difference, then the data collection was biased.

In classification examples, we can use false positive rates and/or false negative rates across sensitive subgroups to see whether these are vastly different. If they are, data is likely to be biased toward one or a couple of subgroups. Another metric that can be used to check whether bias persists is demographic parity, which is a probability comparison of the likelihood of selection from one subgroup over another. If the ratio of probabilities between the higher selection subgroup and the lower selection subgroup is below 0.8, it is quite likely data is biased and does not have enough representative samples. It is recommended to check multiple metrics to understand bias in the data and the algorithm.

To treat such biases, it is recommended, when collecting data, to use techniques such as stratified sampling to ensure that different groups are represented proportionally in the dataset. Now that we have covered types of bias that are easy to identify, in the next section, we will cover some types of bias that are difficult to identify.

Difficult to identify bias

Some types of bias can be challenging because they are biases that individuals may not be consciously aware of. These biases often operate at a subconscious level and can influence perceptions, attitudes, and behaviors. In order to capture these, organizations and individuals need processes and training to ensure these biases are not present in the workspace. Once it has been identified that there was bias in the data collection process or data labeling process, then sensitive labels can be defined to measure and check whether the model is free from bias or whether there is an acceptable level of bias in the model. Some of these biases are described next.

Group attribution bias

This type of bias occurs when attribution is done for the entire data based on some data points. This usually occurs when the data creators have preconceived biases about the types of attributes present in the data. This type of bias can take two forms:

- **In-group bias**: This is a preconceived bias where associated data points resonate with the data creator, hence those data points get a favorable outcome – for example, if a data engineering manager is designing a resume selection algorithm where they believe someone doing a Udacity nanodegree is qualified for the role.

- **Out-group homogeneity bias**: This is a preconceived bias where data points do not resonate with the data creator, hence those data points get a negative outcome – for example, if a data engineering manager is designing a resume selection algorithm where they believe someone not doing a Udacity nanodegree is not qualified for the role.

Let's move on to another type of bias that is difficult to identify.

Implicit bias

This type of bias occurs when data creators make assumptions about the data based on their own mental models and personal experiences. For example, a sentiment analysis model trained on airline food service review data is likely to associate the word "okay" with neutral sentiment. However, some regions of the world use the word "okay" to signify a positive sentiment.

Bias in machine learning can take many forms; hence, we categorize these biases into two main types, **easy to identify** biases and **difficult to identify** biases. Practitioners are known to take a model-centric approach to treat these biases, where modifying the algorithm or using bias-friendly algorithms has been considered acceptable practice. In the next section, we will take an alternative view to the model-centric approach: the data-centric approach.

The data-centric imperative

Addressing bias in machine learning necessitates a holistic approach, with data-centric strategies complementing model-centric techniques. Data-centricity involves taking proactive steps to curate, clean, and enhance the dataset itself, thus minimizing the bias that models can inherit. By embracing data-centric practices, organizations can foster fairness, accountability, and ethical AI.

In the remainder of this chapter, we will explore a spectrum of data-centric strategies that empower machine learning practitioners to reduce bias. These include data resampling, augmentation, cleansing, feature selection, and more. Real-world examples will illustrate the tangible impact of these strategies in the domains of finance, human resources, and healthcare.

If data is fairly and accurately captured or created, then it is quite likely algorithms will be mostly free from bias. However, the techniques we will cover in this chapter are post-data creation, where ML practitioners have to work with provided data.

In the following subsections, we will discuss some data-centric strategies for reducing bias in machine learning without changing the algorithm. These can be referred to as data debiasing techniques.

Sampling methods

Sampling methods such as undersampling and oversampling address class imbalances. Undersampling reduces majority class instances, whereas oversampling augments minority class examples. Integrating both mitigates overfitting and information loss, balancing class representation effectively. These methods can be combined with outlier treatment and Shapley values to further sample the data where harder-to-classify or harder-to-estimate data points can be removed or introduced to enhance the fairness metrics. These techniques are covered next.

Undersampling

In undersampling, we remove random or strategic subsets of overrepresented data points to balance class distributions – deleting data points from overrepresented classes where examples are difficult to classify or at random is a commonly used technique. We can also use outlier removal for regression tasks.

Oversampling

In oversampling, we add random or strategic subsets of underrepresented data points to provide more examples to the algorithm. We can duplicate or generate synthetic data points for underrepresented classes to balance class distributions. We can use techniques such as the **Synthetic Minority Oversampling Technique** (**SMOTE**) and random oversampling for classification tasks. Alternatively, we can utilize outlier or edge case addition/removal for regression tasks.

Combination of undersampling and oversampling

These cover techniques such as SMOTEENN or SMOTETomek, where SMOTE is utilized to oversample the minority class. Techniques such as **Edited Nearest Neighbors** (**ENN**) or Tomek Links are used to remove the examples that are difficult to classify or agree on using nearest neighbors, as these points are close to the boundary and there is no clear separation.

Anomaly detection for oversampling and undersampling the data

This covers using an anomaly detection technique to identify data points that are edge cases, and then these points can be reintroduced multiple times or removed so the model can get a better signal or become more generalized.

Use of Shapley values for oversampling and undersampling data

This covers using Shapley values to oversample or undersample data. Shapley values quantify feature importance by assessing each feature's contribution to a model's prediction. High Shapley values highlight influential features. Removing instances with high Shapley values but wrong predictions might enhance model accuracy by reducing outliers. Oversampling instances with high Shapley values and correct predictions can reinforce the model's understanding of crucial patterns, potentially improving performance.

Other data-centric techniques

Besides sampling methods, there are other data-centric techniques that can be used to reduce bias, some of which have been covered in previous chapters, and some we will utilize in the case study. The three main ones are described next.

Data cleansing

This includes removing missing data, where the inclusion of missing data can lead to unfair outcomes. These techniques were covered in *Chapter 5, Techniques for Data Cleaning*, where missing data was classified as "missing not at random."

Feature selection

This includes selecting specific features or eliminating features that will reduce bias. This may mean identifying a variable that is highly associated with a sensitive variable and outcome label, and removing such indirect variables or removing sensitive variables.

Feature engineering

Feature engineering offers potent tools to mitigate model bias. Techniques such as re-encoding sensitive attributes, creating interaction terms, or introducing proxy variables enable models to learn without direct access to sensitive information. Feature selection and dimensionality reduction methods trim irrelevant or redundant features, fostering fairer and more robust models. Additionally, generating synthetic features or utilizing domain-specific knowledge helps improve models with a better understanding of data, aiding in fairer decision-making while improving overall model performance and reducing bias. We will create a synthetic variable, "Interest," in the example to show how the model is biased toward one subgroup over another.

Now that we have covered data-centric methods, in the next section, we will describe the problem statement, and walk through examples of how we can identify and reduce bias in real life.

Case study

The challenge at hand centers on uncovering and addressing potential bias within a dataset pertaining to credit card defaults in Taiwan. Acquired from the UC Irvine Machine Learning Repository (`https://archive.ics.uci.edu/dataset/350/default+of+credit+card+clients`), this dataset comprises information from 30,000 credit card clients over a six-month span, including demographic factors such as gender, marital status, and education. The key concern is whether these demographic features introduce bias into a decision tree classifier trained on all available features, with a specific focus on gender-related bias. The overarching objective of this example is to not only identify but also mitigate any biased outcomes through the application of data-centric techniques. By reevaluating the algorithm's performance using fairness metrics, the example aims to shed light on the real-world implications of bias in financial decision-making, particularly how these biases can impact individuals based on gender and other demographic factors, potentially leading to unequal treatment in credit assessments and financial opportunities. Addressing and rectifying such biases is crucial for promoting fairness and equity in financial systems.

We will use two key metrics to check the fairness of the algorithm:

- **Equalized odds difference**: This metric compares the false negative rate and false positive rate across the sensitive variables, then takes the maximum difference between the false negative rate and false positive rate. For instance, on the test set, the false positive rate among men and women is 0.3 and 0.2 (difference of 0.1), whereas the false negative rate among men and women is 0.15 and 0.12 (difference of 0.03). Since the difference is larger on the false positive rate, the equalized odds will be 0.1.

- **Demographic parity ratio**: This metric measures whether the predictions made by a model are independent of a sensitive variable, such as race, gender, or age. Given this is a ratio, it measures the ratio of a lower selection rate to that of a higher selection rate. A ratio of 1 means that demographic parity is achieved, whereas below 0.8 usually means that the algorithm is highly biased toward one group of individuals over the others.

The following is a description of the features in the dataset:

- **Independent variables**:

  - `LIMIT_BAL`: Amount of the given credit in NT dollars, including both the individual consumer credit and their family (supplementary) credit.

  - `Sex`: Gender (1 = male; 2 = female).

  - `Education X3`: Education (1 = graduate school; 2 = university; 3 = high school; 4 = others)

  - `Marriage X4`: Marital status (1 = married; 2 = single; 3 = others)

  - `Age X5`: Age of the person in years

  - `PAY_0- PAY_5; X6 - X11`: History of past payments, which includes the past monthly payment records (from April to September 2005), where PAY_0X6 = the repayment status in September, PAY_2; X7 = the repayment status in August 2005; ... PAY_6; X11 = the repayment status in April 2005. The measurement scale for the repayment status is -1 = amount paid duly; 1 = payment delay for 1 month; 2 = payment delay for 2 months; ... ; 8 = payment delay for 8 months; 9 = payment delay for 9 months, and so on.

  - `BILL_AMT1 . BILL_AMT6; X12-X17`: Bill statement amount (in Taiwan dollars). BILL_AMT1;X12 means the amount on the credit card statement as of September 2005, while BILL_AMT6;X17 means the amount on the credit card statement as of April 2005.

  - `PAY_AMT1-PAY_AMT6; X18-X23`: Amount of payments made based on the previous month's bill statement. PAY_AMT1;X18 means amount paid in September 2005, while PAY_AMT6;X23 means amount paid on April 2005.

- **Target variable**:

  - `default payment next month`: Whether a person defaulted on the next month's payment (Yes = 1, No = 0), in 2005

To import the dataset, you need to install `pandas`. We will also use the `os` library to navigate the path and store the dataset. This library is native to Python. We will call the `loan_dataset.csv` file and save it in the same directory, from where we will run this example:

```
import pandas as pd
import os
FILENAME = "./loan_dataset.csv"
DATA_URL = "http://archive.ics.uci.edu/ml/machine-learning-
databases/00350/default%20of%20credit%20card%20clients.xls"
```

The file takes a couple of seconds to a minute based on internet speed, so when we run this example for the first time, the file will be stored locally. However, on the subsequent runs, with the help of the `os` library, we will check that the file exists, else download it. We will rename two variables: `PAY_0` to `PAY_1`, and also `default payment next month` to `default`. We don't believe the `ID` column will be useful for machine learning, hence we will drop it:

```
if not os.path.exists(FILENAME):
    data = (
        pd.read_excel(io=DATA_URL, header=1)
        .drop(columns=["ID"])
        .rename(
            columns={"PAY_0": "PAY_1", "default payment next month":
"default"}
        )
    )
    data.to_csv(FILENAME, sep=",", encoding="utf-8", index=False)
```

Now, we load the file from the local directory into a DataFrame called `dataset`. There are 30,000 rows and 24 columns including the target variable:

```
dataset = pd.read_csv(FILENAME, sep=",", encoding="utf-8")
dataset.shape

(30000, 24)
```

Next, we run the `dataset.info()` method to check whether there are any missing values or wrongly encoded columns:

```
<class 'pandas.core.frame.DataFrame'>
RangeIndex: 30000 entries, 0 to 29999
Data columns (total 24 columns):
 #   Column     Non-Null Count  Dtype
---  ------     --------------  -----
 0   LIMIT_BAL  30000 non-null  int64
 1   SEX        30000 non-null  int64
 2   EDUCATION  30000 non-null  int64
 3   MARRIAGE   30000 non-null  int64
 4   AGE        30000 non-null  int64
 5   PAY_1      30000 non-null  int64
 6   PAY_2      30000 non-null  int64
 7   PAY_3      30000 non-null  int64
 8   PAY_4      30000 non-null  int64
 9   PAY_5      30000 non-null  int64
 10  PAY_6      30000 non-null  int64
 11  BILL_AMT1  30000 non-null  int64
 12  BILL_AMT2  30000 non-null  int64
 13  BILL_AMT3  30000 non-null  int64
 14  BILL_AMT4  30000 non-null  int64
 15  BILL_AMT5  30000 non-null  int64
 16  BILL_AMT6  30000 non-null  int64
 17  PAY_AMT1   30000 non-null  int64
 18  PAY_AMT2   30000 non-null  int64
 19  PAY_AMT3   30000 non-null  int64
 20  PAY_AMT4   30000 non-null  int64
 21  PAY_AMT5   30000 non-null  int64
 22  PAY_AMT6   30000 non-null  int64
 23  default    30000 non-null  int64
dtypes: int64(24)
```

Figure 8.1 – Output of the dataset.info() method

We don't have any missing data; however, three categorical columns (SEX, EDUCATION, and MARRIAGE) have integer data types, which we may have to convert to strings. Since values in SEX might be ordinal, we will first remap them to 1 and 0:

```
cat_colums = ['EDUCATION', 'MARRIAGE']
for col in cat_colums:
    dataset[col] = dataset[col].astype("category")
dataset['SEX'] = dataset['SEX'].map({1: 1, 2:0})
```

If we rerun `dataset.info()`, we will see that the data type for the three columns is now `category`; we can now one-hot encode them. We exclude SEX from one-hot encoding since a person is either a male or female (in this dataset) and that information can be captured in one column. We will also extract SEX and store it in another variable, A, and separate the target variable and independent features. Next, we create a mapping for values in the SEX feature to be used for analysis and visualization, to help interpret the results, so 1 will be mapped to `male` values and 0 will be mapped to `female` values. We store this mapping in the `A_str` variable:

```
Y, A = dataset.loc[:, "default"], dataset.loc[:, "SEX"]
X = pd.get_dummies(dataset.drop(columns=["default","SEX"]))
X["SEX"] = A.copy()

A_str = A.map({1: "male", 0: "female"})
```

Next, let's load all the required libraries.

Loading the libraries

To run the example, you will need the following additional libraries:

- sklearn (scikit-learn) for data preprocessing and fitting the models

- numpy to calculate some metrics and do some data wrangling

- imblearn for over and undersampling

- fairlearn to calculate bias and fairness scores

- shap to visualize the interpretations of the model

We load all the libraries at the start:

```
from sklearn.model_selection import train_test_split, cross_validate
from sklearn.preprocessing import StandardScaler
from sklearn.tree import DecisionTreeClassifier
from imblearn.over_sampling import SMOTE, ADASYN
from imblearn.combine import SMOTEENN, SMOTETomek
from sklearn.ensemble import IsolationForest
from imblearn.pipeline import make_pipeline
from imblearn.under_sampling import AllKNN, InstanceHardnessThreshold,
RepeatedEditedNearestNeighbours, TomekLinks, EditedNearestNeighbours
from sklearn.metrics import balanced_accuracy_score, roc_auc_score,
confusion_matrix, ConfusionMatrixDisplay
from sklearn.pipeline import Pipeline
from fairlearn.metrics import MetricFrame, equalized_odds_difference,
demographic_parity_ratio
import numpy as np
import shap
```

Next, we split the dataset into train and test, using train_test_split, and assign 20% of the data to test. We also split A_str into A_train and A_test, so we can calculate fairness scores on test data:

```
X_train, X_test, y_train, y_test, A_train, A_test = train_test_
split(X,
                Y,
```

```
                    A_str,
                    test_size=0.2,
                    stratify=Y,
                    random_state=42)
```

Next, we create the decision tree classifier pipeline and train the algorithm with the sensitive features:

```
d_tree_params = {
    "min_samples_leaf": 10,
    "random_state": 42
}
estimator = Pipeline(steps=[
    ("classifier", DecisionTreeClassifier(**d_tree_params))
])
estimator.fit(X_train, y_train)
```

Next, we calculate the ROC score and extract the predictions. We also visualize the confusion matrix:

```
y_pred_proba = estimator.predict_proba(X_test)[:, 1]
y_pred = estimator.predict(X_test)
print(f"Roc score is : {roc_auc_score(y_test, y_pred_proba)}")

cm = ConfusionMatrixDisplay(confusion_matrix(y_test, y_pred), display_
labels=estimator.classes_)
cm.plot()

Roc score is : 0.6875636482794665
```

The code generates the following confusion matrix:

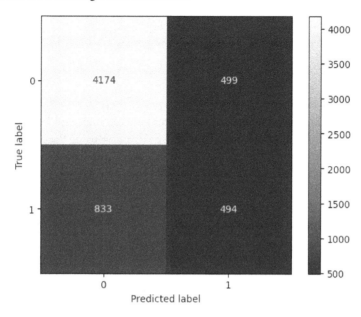

Figure 8.2 – Output confusion matrix

In order to check whether the algorithm is fair or not, we will first calculate false positive and false negative rates, and then compare those across male and female cohorts on the test dataset to see whether there are big differences between the two cohorts.

In the following code block, we have created two functions to calculate a false positive rate and a false negative rate. We have further created a dictionary of fairness metrics, in which we use the false positive rate and false negative rate, alongside a balanced accuracy metric from scikit-learn. We have then created a list of fairness metrics and stored them in a variable for easy access:

```
def false_positive_rate(y_true, y_pred):
    """Compute the standard error for the false positive rate
estimate."""
    tn, fp, fn, tp = confusion_matrix(y_true, y_pred).ravel()
    return fp/(fp+tn)


def false_negative_rate(y_true, y_pred):
    """Compute the standard error for the false positive rate
estimate."""
    tn, fp, fn, tp = confusion_matrix(y_true, y_pred).ravel()
    return fn/(tp+fn)
```

```
fairness_metrics = {
    "balanced_accuracy": balanced_accuracy_score,
    "false_positive_rate": false_positive_rate,
    "false_negative_rate": false_negative_rate,
}


metrics_to_report = list(fairness_metrics.keys())
```

We have also created a function to report the differences between male and female cohorts on the fairness metrics. We first create a DataFrame called metricframe using the convenience function from fairlearn called MetricFrame. It takes in true labels, predictions, and sensitive feature values, along with a dictionary of metrics to report on. We then leverage the .by_group property to report on fairness metrics for each cohort. Within the function, we also report on equalised_odds_difference and demographic_parity_ratio from the fairlearn library to understand the overall fairness of the model:

```
def calculate_fairness_metrics(y_test, y_pred, A_test,
metrics=fairness_metrics):
    """Function to calculate fairness metrics"""
    metricframe = MetricFrame(
        metrics=fairness_metrics,
        y_true=y_test,
        y_pred=y_pred,
        sensitive_features=A_test,
    )

    print(metricframe.by_group[metrics_to_report])
    print("\n *diff*")
    print(metricframe.difference()[metrics_to_report])
    print("\n *final_metrics*")
    print(metricframe.overall[metrics_to_report])

    equalized_odds = equalized_odds_difference(
        y_test, y_pred, sensitive_features=A_test
    )
    print("\n *equalized_odds*")
    print(equalized_odds)

    dpr= demographic_parity_ratio(y_test, y_pred, sensitive_
features=A_test)
    print("\n *demographic_parity_ratio*")
    print(dpr)
```

We now run the function and calculate the fairness scores. It is evident that the model is quite similar in male and female cohorts since false positive rates and false negative rates are similar among the cohorts. Since the difference in the false positive rate is larger than the false negative rate, the equalized odds difference is the same as the difference between the false positive rate of the two groups. We can also see that the demographic parity ratio is above 0.8, which means that both cohorts are quite likely to get selected for a positive outcome:

```
calculate_fairness_metrics_unmitigated = calculate_fairness_metrics(y_
test, y_pred, A_test)
```

This will display the following output:

```
              balanced_accuracy  false_positive_rate  false_negative_rate
SEX
female             0.634938              0.099576             0.630548
male               0.629122              0.117871             0.623886

  *diff*
balanced_accuracy    0.005816
false_positive_rate  0.018294
false_negative_rate  0.006662
dtype: float64

  *final_metrics*
balanced_accuracy    0.632742
false_positive_rate  0.106784
false_negative_rate  0.627732
dtype: float64

  *equalized_odds*
0.018294451247019394

  *demographic_parity_ratio*
0.8812852934911919
```

Figure 8.3 – Fairness scores

To illustrate bias in the dataset, we may need to generate a synthetic variable that correlates with a real-world scenario where, based on history, a cohort is treated more unfairly. First, we compare the default rate across males and females in the training dataset. We then add the synthetic noise:

```
for val in dataset.SEX.unique():
    print(f"{('male' if val == 1 else 'female')} default rate is: ")
    print(dataset[dataset.SEX == val]['default'].mean())
    print()

female default rate is:
0.20776280918727916

male default rate is: 0.2416722745625841
```

Given that the male default is higher than the female, we can replicate a biased scenario where applicants with lower default rates will have lower interest rates, but applicants with higher default rates will have higher interest rates imposed by the bank. Let's assume the bank managers believe males are more likely to default, and instead of generalizing the scenario, the bank decides to charge higher interest rates to males.

To mimic this scenario, we will introduce a new feature, `Interest_rate`, following a Gaussian distribution. The mean will be 0 where someone hasn't defaulted, but will be 2 times 1 where someone has defaulted. We also set the standard deviation to 2 for males and 1 for females.

To generate the synthetic Gaussian distribution, we use the `numpy.random.normal` method, with a seed of 42 for reproducibility:

```
np.random.seed(42)
X.loc[:, 'Interest_rate'] = np.random.normal(loc=2*Y, scale=A.
map({1:2, 0:1}))

print("Maximum interest rate for men who defaulted vs women who
defaulted")
print(X[(X.SEX == 1) & (Y == 1)]["Interest_rate"].max(), X[(X.SEX ==
0) & (Y == 1)]["Interest_rate"].max())


print()


print("Maximum interest rate for men who did not default vs women that
did not default")
print(X[(X.SEX == 1) & (Y == 0)]["Interest_rate"].max(), X[(X.SEX ==
0) & (Y == 0)]["Interest_rate"].max())

Maximum interest rate for men who defaulted vs women who defaulted
9.852475412872653 6.479084251025757

Maximum interest rate for men who did not default vs women that did
not default
6.857820956016427 3.852731490654721
```

Now that we have added the noise, we retrain the algorithm with the interest variable and recalculate the fairness metrics. We first split the data, then retrain and recalculate the fairness metrics. We resplit the data into `train` and `test`, as shown previously, and retrain the algorithm. Once retrained, we calculate the impact.

We can see, in the following code, that by adding the synthetic interest variable, we have improved the ROC metric:

```
y_pred_proba = estimator.predict_proba(X_test)[:, 1]
y_pred = estimator.predict(X_test)
```

```
roc_auc_score(y_test, y_pred_proba)
```

```
0.8465698909107798
```

It is clear from the following output that we now have a more biased algorithm, based on equalized odds. The false negative rate is quite high in males, which means that more males who are unlikely to pay back to the bank are likely to be given a loan, and if this model was productionized, this could result in unfair outcomes:

```
calculate_fairness_metrics(y_test, y_pred, A_test)
```

This will print the following information:

```
              balanced_accuracy  false_positive_rate  false_negative_rate
SEX
female              0.813874            0.062853             0.309399
male                0.696357            0.113525             0.493761

  *diff*
balanced_accuracy        0.117517
false_positive_rate      0.050672
false_negative_rate      0.184362
dtype: float64

  *final_metrics*
balanced_accuracy        0.764922
false_positive_rate      0.082816
false_negative_rate      0.387340
dtype: float64

  *equalized_odds*
0.1843616630131758

  *demographic_parity_ratio*
0.9573799733230204
```

Figure 8.4 – Fairness scores

To reduce the bias, we will apply the first data-centric debiasing technique under the feature selection by removing the sensitive variable from the algorithm. This can be done by retraining the algorithm without the SEX variable.

Given that the dataset is biased toward one gender due to a higher variation in interest rates, it is recommended in the real world that data engineers and data scientists work with domain experts and data producers to reduce this bias in the dataset. For instance, instead of using SEX to determine the interest rate, other features could be used, such as payment history, credit history, and income. In the training step, we can drop the SEX variable:

```
estimator.fit(X_train.drop(['SEX'], axis=1), y_train)
Pipeline(steps=[('classifier',
                 DecisionTreeClassifier(min_samples_leaf=10, random_
state=42))])
```

From the following output, we can see that by removing the SEX variable, the ROC score has dropped from 0.846 to 0.839:

```
estimator.fit(X_train.drop(['SEX'], axis=1), y_train)
y_pred_proba = estimator.predict_proba(X_test.drop(['SEX'], axis=1))
[:, 1]
y_pred = estimator.predict(X_test.drop(['SEX'], axis=1))
roc_auc_score(y_test, y_pred_proba)

0.8392395442658211
```

Looking at the following fairness metrics, it is obvious that when the outcome is biased based on the cohort of data, removing the variable from the training can debias the algorithm. The false negative rate in male has decreased, whereas, in female, it has increased; however, the algorithm is more fair compared to when the SEX variable was used. The equalized odds have dropped from 0.18 to 0.07, but the demographic parity ratio has reduced, which means one group has more chance of getting a loan than the other:

```
calculate_fairness_metrics_mitigated_v1 = calculate_fairness_
metrics(y_test, y_pred, A_test)
```

The output is as follows:

```
          balanced_accuracy  false_positive_rate  false_negative_rate
SEX
female             0.777149             0.050141             0.395561
male               0.720184             0.121130             0.438503

  *diff*
balanced_accuracy     0.056965
false_positive_rate   0.070989
false_negative_rate   0.042941
dtype: float64

  *final_metrics*
balanced_accuracy     0.754088
false_positive_rate   0.078108
false_negative_rate   0.413715
dtype: float64

  *equalized_odds*
0.0709885778117395

  *demographic_parity_ratio*
0.75073202584666
```

Figure 8.5 – Fairness metrics

Next, we will show you how to apply undersampling techniques to ensure the outcome variable is balanced.

AllKNN undersampling method

We will start with the AllKNN algorithm from the `imblearn` package, and then try the instant hardness algorithm. Since the algorithms use KNN under the hood, which is a distance-based measure, we need to ensure that we scale the features using the scikit-learn `StandardScaler` method. We will first scale the variables, then run the sampling algorithm, and then train the decision tree. We will run the algorithm with 5 k cross-validation, and ensure the function returns the model trained. Cross-validation will be scored on `roc_auc` and balanced accuracy.

We will first try an undersampling technique, `AllKNN`, from `imblearn`. This algorithm does not aim at balancing majority and minority classes; however, it removes instances that are harder to classify from the majority class. It does that iteratively where, first, the model is trained on the entire dataset. Then, in the prediction step of the majority class, if a disagreement occurs between any of the neighbors about the predicted outcome, the data point is removed from the majority class. In the first iteration, a 1-KNN model is trained and some samples are removed, and then in the next iteration, a 2-KNN model is trained, and in the following iteration, a 3-KNN model is trained. Usually, the algorithm (by default) will end at the 3-KNN iteration; however, the practitioner can choose more iterations, and the algorithm will not stop until the number of samples between the majority and minority class becomes the same or a maximum number of iterations is reached – whichever happens earlier.

Let's first define the scaler and sampler method:

```
scaler = StandardScaler()
sampler_method = AllKNN(n_jobs=-1)
```

Next, we create a pipeline object and pass the scaler, sampler, and estimator to the pipeline:

```
sampler_pipeline = make_pipeline(
    scaler,
    sampler_method,
    estimator)
```

Then we pass the training data and run the cross-validation. We set the cross-validation method to return the estimator (pipeline) by setting `return_estimator=True`, so that we can use it to make predictions on the test data:

```
cv_results = cross_validate(sampler_pipeline,
                            X_train.drop(['SEX'], axis=1),
                            y_train, scoring=['roc_auc','balanced_
accuracy'],
                            return_estimator=True)
```

Next, we print the mean and standard deviation of ROC and balanced accuracy from the cross-validation step, returned from prediction results in each step, where at each step, four folds were used on training and the prediction was made on the fifth fold:

```
print(f"Validation roc auc : {cv_results['test_roc_auc'].mean():.3f}
+/- {cv_results['test_roc_auc'].std():.3f}")
print(f"Validation balanced acc : {cv_results['test_balanced_
accuracy'].mean():.3f} +/- {cv_results['test_balanced_accuracy'].
std():.3f}")


Validation roc auc : 0.853 +/- 0.006
Validation balanced acc : 0.802 +/- 0.005
```

We can see that by removing hard examples using the undersampling technique, `roc_auc` on `test` data bumped from 0.839 in the previous step to 0.85:

```
model = sampler_pipeline.fit( X_train.drop(['SEX'], axis=1), y_train)
y_pred_proba = model.predict_proba(X_test.drop(['SEX'],axis=1))[:, 1]
y_pred = model.predict(X_test.drop(['SEX'],axis=1))
roc_auc_score(y_test, y_pred_proba)
0.8537904984477683
```

Next, we calculate the fairness metrics. Although the false negative rate has decreased for both males and females, the false positive rate has increased, and the equalized odds difference has also increased from the previous step. This might be because cases with male samples that were difficult to classify have been removed:

```
calculate_fairness_metrics_mitigated_v2 = calculate_fairness_
metrics(y_test, y_pred, A_test)
```

```
                balanced_accuracy  false_positive_rate  false_negative_rate
SEX
female              0.839760             0.137712             0.182768
male                0.744553             0.254210             0.256684

*diff*
balanced_accuracy     0.095207
false_positive_rate   0.116498
false_negative_rate   0.073917
dtype: float64

*final_metrics*
balanced_accuracy     0.801188
false_positive_rate   0.183608
false_negative_rate   0.214017
dtype: float64

*equalized_odds*
0.11649780425155823

*demographic_parity_ratio*
0.7664119740094152
```

Figure 8.6 – Fairness metrics

We will now explore the impact on the fairness metrics by introducing hard cases.

Instance hardness undersampling method

As the name suggests, the instance hardness method focuses on samples that are harder to classify, which are usually at the boundary or overlap with other classes. Usually, this depends on the algorithm used (as some algorithms are better at some hard cases than others) and the level of overlap between the classes. For such samples, the learning algorithm will usually show the low probability prediction on the hard cases, which means the lower the probability, the higher the instance hardness. Under the hood, the method has the capability to retain the right number of samples, based on the class imbalance.

In the first step, we will define the algorithm, and the algorithm will be passed on to the instance hardness step. We will then define the instance hardness undersampling method, with three-fold cross-validation.

Next, we create the decision tree estimator. Finally, we combine the steps in the pipeline with scaling the dataset, then undersampling the data, and finally, training the model. When the pipeline is defined, we run the cross-validation similar to the previous pipeline:

```
d_tree_params = {
    "min_samples_leaf": 10,
    "random_state": 42
}


d_tree = DecisionTreeClassifier(**d_tree_params)

sampler_method = InstanceHardnessThreshold(
    estimator=d_tree,
    sampling_strategy='auto',
    random_state=42,
    n_jobs=-1,
    cv=3)

estimator = Pipeline(steps=[
    ("classifier", DecisionTreeClassifier(**d_tree_params))
])
sampler_pipeline = make_pipeline(
    scaler,
    sampler_method,
    estimator)

cv_results = cross_validate(sampler_pipeline, X_train.drop(['SEX'],
axis=1), y_train, scoring=['roc_auc','balanced_accuracy'], return_
estimator=True)
```

Both `AllKNN` and `InstanceHardness` returned similar cross-validation results:

```
print(f"Validation roc auc : {cv_results['test_roc_auc'].mean():.3f}
+/- {cv_results['test_roc_auc'].std():.3f}")
print(f"Validation balanced acc : {cv_results['test_balanced_
accuracy'].mean():.3f} +/- {cv_results['test_balanced_accuracy'].
std():.3f}")
Validation roc auc : 0.853 +/- 0.005
Validation balanced acc : 0.807 +/- 0.007
```

The ROC slightly bumped from 0.85 to 0.854 on the `test` data when using the instance hardness method:

```
model = sampler_pipeline.fit( X_train.drop(['SEX'], axis=1), y_train)
y_pred_proba = model.predict_proba(X_test.drop(['SEX'],axis=1))[:, 1]
y_pred = model.predict(X_test.drop(['SEX'],axis=1))
roc_auc_score(y_test, y_pred_proba)


0.8549627959428299
```

The fairness metrics are quite similar to the previous undersampling technique, and probably due to similar reasons, where, by removing difficult cases, the model is unable to deal with predicting difficult `male` cases. However, in both undersampling methods, the equalized odds have increased, compared to the feature selection step. Also, the demographic parity ratio is still under 0.8, which means one subclass of gender is more likely to be selected over another when predicting `default`:

```
calculate_fairness_metrics_mitigated_v3 = calculate_fairness_
metrics(y_test, y_pred, A_test)


                balanced_accuracy  false_positive_rate  false_negative_rate
SEX
female                 0.844083             0.135593             0.176240
male                   0.742652             0.258012             0.256684

 *diff*
balanced_accuracy       0.101432
false_positive_rate     0.122419
false_negative_rate     0.080444
dtype: float64

 *final_metrics*
balanced_accuracy       0.802965
false_positive_rate     0.183822
false_negative_rate     0.210249
dtype: float64

 *equalized_odds*
0.1224187296881761

 *demographic_parity_ratio*
0.7596491075011778
```

Figure 8.7 – Fairness metrics

Next, let's look at oversampling methods.

Oversampling methods

Another way of improving model performance and fairness metrics is by introducing additional examples. The next two oversampling techniques, SMOTE and ADASYN, were introduced in *Chapter 7*, *Using Synthetic Data in Data-Centric Machine Learning*, hence we will not cover the details behind the algorithm. We will use these techniques in the context of improving fairness metrics by adding additional examples, in the hope that the model is able to learn better with additional data points.

For each of the methods, we will first scale the dataset, add additional minority class examples, and then train the model. We will print the cross-validation scores and the test ROC score, as well as fairness metrics.

SMOTE

Given that we used this algorithm in *Chapter 7*, *Using Synthetic Data in Data-Centric Machine Learning*, we will dive straight into the code:

```
estimator = Pipeline(steps=[
    ("classifier", DecisionTreeClassifier(**d_tree_params))
])
sampler_method = SMOTE(random_state=42)

sampler_pipeline = make_pipeline(
    scaler,
    sampler_method,
    estimator)

cv_results = cross_validate(sampler_pipeline, X_train.drop(['SEX'],
axis=1), y_train, scoring=['roc_auc','balanced_accuracy'], return_
estimator=True)

print(f"Validation roc auc : {cv_results['test_roc_auc'].mean():.3f}
+/- {cv_results['test_roc_auc'].std():.3f}")
print(f"Validation balanced acc : {cv_results['test_balanced_
accuracy'].mean():.3f} +/- {cv_results['test_balanced_accuracy'].
std():.3f}")

model = sampler_pipeline.fit( X_train.drop(['SEX'], axis=1), y_train)
y_pred_proba = model.predict_proba(X_test.drop(['SEX'],axis=1))[:, 1]
y_pred = model.predict(X_test.drop(['SEX'],axis=1))
roc_auc_score(y_test, y_pred_proba)

Validation roc auc : 0.829 +/- 0.009
```

```
Validation balanced acc : 0.758 +/- 0.012


0.8393191272926885
```

The validation metrics and the `test` ROC score show poorer results compared to the undersampling methods covered previously. In the next step, we explore the fairness metrics:

```
calculate_fairness_metrics_mitigated_v4 = calculate_fairness_
metrics(y_test, y_pred, A_test)
```

	balanced_accuracy	false_positive_rate	false_negative_rate
SEX			
female	0.802982	0.126412	0.267624
male	0.710597	0.208039	0.370766

```
 *diff*
balanced_accuracy     0.092385
false_positive_rate   0.081627
false_negative_rate   0.103142
dtype: float64

 *final_metrics*
balanced_accuracy     0.765101
false_positive_rate   0.158571
false_negative_rate   0.311228
dtype: float64

 *equalized_odds*
0.10314246752581879

 *demographic_parity_ratio*
0.833584328249994
```

Figure 8.8 – Fairness metrics

The fairness metrics are better in comparison with the undersampling methods – that is, the difference between false positive and false negative rates is reduced between men and women and, based on the demographic parity ratio, the model is more likely to select both types of gender applicants for loan default. In the next section, we will use the ADASYN algorithm and compare it with SMOTE and other undersampling methods.

ADASYN

Similarly to the SMOTE method, we covered ADASYN in *Chapter 7, Using Synthetic Data in Data-Centric Machine Learning,* hence we will dive straight into the code, in which we oversample the minority class:

```
sampler_method = ADASYN(random_state=42)

sampler_pipeline = make_pipeline(
    scaler,
    sampler_method,
    estimator)

cv_results = cross_validate(sampler_pipeline, X_train.drop(['SEX'],
axis=1), y_train, scoring=['roc_auc','balanced_accuracy'], return_
estimator=True)

print(f"Validation roc auc : {cv_results['test_roc_auc'].mean():.3f}
+/- {cv_results['test_roc_auc'].std():.3f}")
print(f"Validation balanced acc : {cv_results['test_balanced_
accuracy'].mean():.3f} +/- {cv_results['test_balanced_accuracy'].
std():.3f}")

model = sampler_pipeline.fit( X_train.drop(['SEX'], axis=1), y_train)
y_pred_proba = model.predict_proba(X_test.drop(['SEX'],axis=1))[:, 1]
y_pred = model.predict(X_test.drop(['SEX'],axis=1))
roc_auc_score(y_test, y_pred_proba)

Validation roc auc : 0.823 +/- 0.004
Validation balanced acc : 0.757 +/- 0.006

0.816654655300673
```

The validation metrics and `test` ROC score are slightly below the SMOTE results and undersampling methods. Now, let's review the fairness metrics:

```
calculate_fairness_metrics_mitigated_v5 = calculate_fairness_
metrics(y_test, y_pred, A_test)
```

```
              balanced_accuracy   false_positive_rate   false_negative_rate
SEX
female              0.790323            0.150424              0.268930
male                0.691245            0.232482              0.385027

 *diff*
balanced_accuracy      0.099078
false_positive_rate    0.082059
false_negative_rate    0.116097
dtype: float64

 *final_metrics*
balanced_accuracy      0.749619
false_positive_rate    0.182752
false_negative_rate    0.318011
dtype: float64

 *equalized_odds*
0.11609723405146533

 *demographic_parity_ratio*
0.8515482584474486
```

Figure 8.9 – Fairness metrics

The equalized odds are slightly higher for ADASYN, whereas demographic parity is slightly better when compared to SMOTE, and both oversampling techniques guarantee higher fairness over undersampling methods, but slightly poorer ROC performance.

We have now seen that, despite balancing the classes, model fairness is compromised, and it is mostly male examples where the model is making more errors. So, in the next section, we will randomly introduce some additional male examples where the model is misclassifying positive cases.

Oversampling plus misclassified examples at random

We will first balance the dataset using ADASYN and avoid undersampling techniques since we want to retain hard cases that are difficult to classify. We then train the model and identify male cases that the model believes should be positive but wrongly classifies as negative. We then randomly select 10% of these cases, add them back to the training dataset, and retrain the model with the same algorithm.

At the end, we review the model metrics and fairness metrics on the test data.

We utilize oversampling and reintroduce misclassified data points at random. Let's run the pipeline with the ADASYN oversampling method:

```
sampler_method = ADASYN(random_state=42)
sampler_pipeline = make_pipeline(
    scaler,
    sampler_method,
    estimator)
```

```
cv_results = cross_validate(sampler_pipeline, X_train.drop(['SEX'],
axis=1), y_train, scoring=['roc_auc','balanced_accuracy'], return_
estimator=True)
model = sampler_pipeline.fit( X_train.drop(['SEX'], axis=1), y_train)
```

Next, we identify the examples from the training dataset where the model is making errors on the male population – that is, examples where the model predicts false negatives. We first subset the data associated with males and then run predictions over this data:

```
X_train_males = X_train[X_train.SEX == 1].copy()
X_train_males["predictions"] = model.predict(X_train_males.
drop(['SEX'], axis=1))
X_train_males['y_true'] = y_train.filter(X_train_males.index)
```

Then we subset this data where the true label is 1 but model predictions are 0:

```
X_train_male_false_negatives = X_train_males[(X_train_males.y_true ==
1) & (X_train_males.predictions == 0)]
```

We randomly select 10% of the values and add them to the X_train dataset. We leverage the .sample method, and this random selection is done with replacement:

```
X_train_sample = X_train_male_false_negatives[X_train.columns].
sample(frac=0.1, replace=True, random_state=42, axis=0)
y_train_sample = X_train_male_false_negatives['y_true'].
sample(frac=0.1, replace=True, random_state=42, axis=0)
```

Next, we add this 10% to X_train and y_train and create a new dataset:

```
X_train_with_male_samples = pd.concat([X_train, X_train_sample],
axis=0, ignore_index=True)
y_train_with_male_samples = pd.concat([y_train, y_train_sample],
axis=0, ignore_index=True)
```

Then, we train the algorithm on this new dataset and print out the validation metrics and the test ROC score:

```
cv_results = cross_validate(sampler_pipeline, X_train_with_
male_samples.drop(['SEX'], axis=1), y_train_with_male_samples,
scoring=['roc_auc','balanced_accuracy'], return_estimator=True)

print(f"Validation roc auc : {cv_results['test_roc_auc'].mean():.3f}
+/- {cv_results['test_roc_auc'].std():.3f}")
print(f"Validation balanced acc : {cv_results['test_balanced_
accuracy'].mean():.3f} +/- {cv_results['test_balanced_accuracy'].
std():.3f}")

model = sampler_pipeline.fit(X_train_with_male_samples.drop(['SEX'],
```

```
axis=1), y_train_with_male_samples)
y_pred_proba = model.predict_proba(X_test.drop(['SEX'],axis=1))[:, 1]
y_pred = model.predict(X_test.drop(['SEX'],axis=1))
roc_auc_score(y_test, y_pred_proba)

Validation roc auc : 0.824 +/- 0.005
Validation balanced acc : 0.754 +/- 0.005
0.8201623558253082
```

Compared to the oversampling section, the validation metrics are quite similar, as is the `test` ROC score. Next, we review the fairness metrics to check whether they have improved:

```
calculate_fairness_metrics_mitigated_v6 = calculate_fairness_
metrics(y_test, y_pred, A_test)
```

```
            balanced_accuracy  false_positive_rate  false_negative_rate
SEX
female              0.784711             0.156427             0.274151
male                0.698027             0.231396             0.372549

 *diff*
balanced_accuracy       0.086684
false_positive_rate     0.074969
false_negative_rate     0.098398
dtype: float64

 *final_metrics*
balanced_accuracy       0.749144
false_positive_rate     0.185962
false_negative_rate     0.315750
dtype: float64

 *equalized_odds*
0.09839758357651152

 *demographic_parity_ratio*
0.8572307380135494
```

Figure 8.10 – Fairness metrics

By adding some false negative `male` examples, we can see that the equalized odds have improved slightly to 0.098 and the demographic ratio has also improved, increasing to 0.85. We believe that even better results can be achieved if we add false positive examples and false negative examples and combine these with undersampling and oversampling techniques.

To demonstrate this, we will iterate over four undersampling techniques (AllKNN, RepeatedEditedNearestNeighbours, InstanceHardnessThreshold, and Tomek), two oversampling techniques (SMOTE and ADASYN), and two combinations of over and undersampling techniques (SMOTEENN and SMOTETomek). How these algorithms work is outside the scope of this example. Instead, the goal is to demonstrate how these data techniques can lead to better selection and a generalized model with slightly poorer performance, but higher fairness.

We will now develop a mechanism where we first train the algorithm and then add false positive examples and false negative examples. Once the examples are added, we run the pipeline by sampling the dataset, using the previous algorithms. We'll record fairness outcomes and the ROC score to find the technique that best fosters a balance between fairness and performance in our algorithm.

We will first create a dictionary with a configuration of each of the aforementioned sampling techniques so we can iterate over it. We can call this AutoML for the sampling technique:

```
methods = {
    "all_knn": AllKNN(n_jobs=-1),
    "renn": RepeatedEditedNearestNeighbours(n_jobs=-1),
    "iht": InstanceHardnessThreshold(
        estimator=DecisionTreeClassifier(**d_tree_params),
        random_state=42,
        n_jobs=-1,
        cv=3),
    "tomek": TomekLinks(n_jobs=-1),
    "adasyn" : ADASYN(random_state=42),
    "smote" : SMOTE(random_state=42),
    "smoteenn": SMOTEENN(random_state=42,
                        smote=SMOTE(random_state=42),
                        enn=EditedNearestNeighbours(n_jobs=-1)
                        ),
    "smotetomek": SMOTETomek(random_state=42,
                            smote=SMOTE(random_state=42),
                            tomek=TomekLinks(n_jobs=-1)
                            )
        }
```

Next, we create two functions that take the training dataset, model, column, and its subset value to help create random samples. The following function will sample false positives:

```
def sample_false_positives(X_train, y_train, estimator, perc=0.1,
subset_col="SEX", subset_col_value=1, with_replace=True):
    """Function to sample false positives"""

    X_train = X_train.copy()
    y_train = y_train.copy()
```

```
    X_train_subset = X_train[X_train[subset_col] == subset_col_value].
copy()
    y_train_subset = y_train.filter(X_train_subset.index).copy()
    X_train_subset["predictions"] = estimator.predict(X_train_subset.
drop([subset_col], axis=1))
    X_train_subset['y_true'] = y_train_subset.values
    X_train_subset_false_positives = X_train_subset[(X_train_subset.y_
true == 0) & (X_train_subset.predictions == 1)]

    X_train_sample = X_train_subset_false_positives[X_train.columns].
sample(frac=perc, replace=with_replace, random_state=42, axis=0)
    y_train_sample = X_train_subset_false_positives['y_true'].
sample(frac=perc, replace=with_replace, random_state=42, axis=0)

    X_train_sample = pd.concat([X_train, X_train_sample], axis=0,
ignore_index=True)
    y_train_sample = pd.concat([y_train, y_train_sample], axis=0,
ignore_index=True)

    return X_train_sample, y_train_sample
```

And this function samples false negatives. By default, both methods will add 10% random examples with replacement:

```
def sample_false_negatives(X_train, y_train, estimator, perc=0.1,
subset_col="SEX", subset_col_value=1, with_replace=True):
    """Function to sample false positives"""

    X_train = X_train.copy()
    y_train = y_train.copy()
    X_train_subset = X_train[X_train[subset_col] == subset_col_value].
copy()
    y_train_subset = y_train.filter(X_train_subset.index).copy()
    X_train_subset["predictions"] = estimator.predict(X_train_subset.
drop([subset_col], axis=1))
    X_train_subset['y_true'] = y_train_subset.values
    X_train_subset_false_negatives = X_train_subset[(X_train_subset.y_
true == 1) & (X_train_subset.predictions == 0)]

    X_train_sample = X_train_subset_false_negatives[X_train.columns].
sample(frac=perc, replace=with_replace, random_state=42, axis=0)
    y_train_sample = X_train_subset_false_negatives['y_true'].
sample(frac=perc, replace=with_replace, random_state=42, axis=0)

    X_train_sample = pd.concat([X_train, X_train_sample], axis=0,
```

```
ignore_index=True)
    y_train_sample = pd.concat([y_train, y_train_sample], axis=0,
ignore_index=True)

    return X_train_sample, y_train_sample
```

Next, we create a function that calculates test metrics post-data improvements. The function takes the test data and estimator and returns model metrics and fairness metrics:

```
def calculate_metrics(estimator, X_test, y_test, A_test):
    """Function to calculate metrics"""

    y_pred_proba = estimator.predict_proba(X_test)[:, 1]
    y_pred = model.predict(X_test)
    roc_auc = roc_auc_score(y_test, y_pred_proba)

    balanced_accuracy = balanced_accuracy_score(y_test, y_pred)

    equalized_odds = equalized_odds_difference(
        y_test, y_pred, sensitive_features=A_test
    )

    dpr = demographic_parity_ratio(y_test, y_pred, sensitive_
features=A_test)

    return roc_auc, balanced_accuracy, equalized_odds, dpr
```

Next, we create a pipeline that will sample the dataset and then create random false positive `male` and false negative `male` examples. We then combine these into the training data, one at a time, and retrain the same algorithm. We then calculate the metrics and store them in a list called `results` with columns. Each iteration adds false negative and false positive examples with model performance and fairness metrics. We then use this list to compare the results across algorithms.

Note

The code for creating the pipeline is pretty lengthy. Please refer to GitHub for the full code: https://github.com/PacktPublishing/Data-Centric-Machine-Learning-with-Python/tree/main/Chapter%208%20-%20Techniques%20for%20identifying%20and%20removing%20bias

---

Next, we create a DataFrame called `df` and add all the `test` metrics so we can compare which method reaped the best model performance and fairness metrics:

```
df = pd.DataFrame(data=results,
                  columns=["method", "sample",
```

```
                                "test_roc_auc", "test_balanced_accuracy",
                                "equalized_odds",
                                "demographic_parity_ratio",
                                "validation_roc_auc",
                                "validation_balanced_accuracy"]
                        )
```

Let's sort the DataFrame based on equalized odds:

```
df.sort_values(by="equalized_odds")
```

We can see that when the dataset was sampled with Tomek Links, where difficult cases were removed from the boundary and combined with additional false positive `male` training samples, this resulted in the best equalized odds of 0.075; however, a demographic parity of 0.8 was not achieved. When the SMOTETomek technique was used in combination with false negative `male` examples, the model achieved a 0.088 equalized odds ratio, which was the best among the sampling methods, and the model also achieved a high demographic parity ratio.

	method	sample	test_roc_auc	test_balanced_accuracy	equalized_odds	demographic_parity_ratio	validation_roc_auc	validation_balanced_accuracy
6	tomek	fp_samples	0.840028	0.755548	0.075251	0.766202	0.837717	0.753923
7	tomek	fn_samples	0.846239	0.762665	0.085028	0.762807	0.838964	0.752152
15	smotetomek	fn_samples	0.831481	0.759142	0.088305	0.889503	0.833660	0.759392
9	adasyn	fn_samples	0.820162	0.749144	0.098398	0.857231	0.824482	0.754271
11	smote	fn_samples	0.827696	0.759188	0.099829	0.842699	0.834220	0.762578
14	smotetomek	fp_samples	0.823669	0.758490	0.107787	0.830200	0.831226	0.767442
8	adasyn	fp_samples	0.820265	0.746181	0.108264	0.900495	0.822120	0.764570
2	renn	fp_samples	0.855651	0.808500	0.108267	0.801404	0.850112	0.803545
1	all_knn	fn_samples	0.854233	0.794824	0.110385	0.773968	0.854162	0.803953
0	all_knn	fp_samples	0.853206	0.796750	0.110388	0.763074	0.848643	0.797514
12	smoteenn	fp_samples	0.838054	0.765218	0.113160	0.850041	0.832612	0.761145
13	smoteenn	fn_samples	0.827099	0.759822	0.113989	0.866365	0.831088	0.758094
10	smote	fp_samples	0.822697	0.752652	0.116223	0.817015	0.828602	0.759015
4	iht	fp_samples	0.856844	0.808137	0.116905	0.774378	0.851046	0.803981
5	iht	fn_samples	0.862331	0.802341	0.118480	0.784571	0.854395	0.801472
3	renn	fn_samples	0.849737	0.808104	0.125404	0.765626	0.854015	0.799935

Figure 8.11 – Resulting output dataset

Oversampling with anomalies

In the previous steps, we learned that by adding poorly classified examples to the training dataset, we were able to improve model fairness. In the next step, instead of choosing samples at random, we will utilize an algorithm that identifies anomalies and then we add these anomalies to the training dataset as an oversampling mechanism.

First, we create a pipeline to oversample the minority class:

```
X_train_scaled = pd.DataFrame()
scaler = StandardScaler()
```

```
sampler = SMOTETomek(random_state=42,
                     smote=SMOTE(random_state=42),
                     tomek=TomekLinks(n_jobs=-1)
                    )
```

Next, we extract the oversampled data. There is no reason why undersampling or no sampling could not have been chosen. Once the oversampled data is extracted, we then scale it back to the original feature space:

```
columns = X_train.drop(['SEX'], axis=1).columns
X_train_scaled[columns] = scaler.fit_transform(X_train.drop(['SEX'],
axis=1))
X_train_resample, y_train_resample = sampler.fit_resample(X_train_
scaled, y_train)
X_train_resample[columns] = scaler.inverse_transform(X_train_resample)
```

Next, we train the isolation forest to identify 10% of anomalies. To do that, we set the contamination to 0.1. We then fit the model on resampled data, and run prediction on this data. We store the results in a column called IF_anomaly and add it to the resampled dataset. We then extract these anomalies, as isolation forest labels with a value of -1:

```
anomaly_model = IsolationForest(contamination=float(.1), random_
state=42, n_jobs=-1)
anomaly_model.fit(X_train_resample)
X_train_resample['IF_anomaly'] = anomaly_model.predict(X_train_
resample)
X_train_resample['default'] = y_train_resample
X_train_additional_samples = X_train_resample[X_train_resample.IF_
anomaly == -1]
X_train_additional_samples.drop(['IF_anomaly'], axis=1, inplace=True)
```

Next, we add these additional data points to the original dataset and train the decision tree model. Once the model is fitted, we calculate the ROC score on the test data. We can see that this is 0.82:

```
X_train_clean = X_train_resample[X_train_resample.IF_anomaly != -1]
y_train_clean = X_train_clean.default

estimator.fit(X_train_clean.drop(['IF_anomaly', 'default'], axis=1),
y_train_clean)
y_pred_proba = estimator.predict_proba(X_test.drop(['SEX'], axis=1))
[:, 1]
y_pred = estimator.predict(X_test.drop(['SEX'], axis=1))

roc_auc_score(y_test, y_pred_proba)
0.8248481592937735
```

Next, we calculate the fairness metrics. Based on the following results, we can say that the model trained in the previous section produced better fairness and demographic parity ratio scores:

```
            balanced_accuracy  false_positive_rate  false_negative_rate
SEX
female               0.794368             0.138418             0.272846
male                 0.701329             0.219446             0.377897

 *diff*
balanced_accuracy    0.093039
false_positive_rate  0.081028
false_negative_rate  0.105051
dtype: float64

 *final_metrics*
balanced_accuracy    0.756201
false_positive_rate  0.170340
false_negative_rate  0.317257
dtype: float64

 *equalized_odds*
0.10505066018811993

 *demographic_parity_ratio*
0.8413623622980747
```

Figure 8.12 – Fairness metrics

Now that we have utilized various examples of undersampling and oversampling data, including reintroducing random misclassified examples and anomalies, in the next section, we will utilize an advanced technique, where we will be more selective with which examples to add and which examples to remove, to further reduce bias in the algorithm.

Shapley values to detect bias, oversample, and undersample data

In this section, we will utilize Shapley values to identify examples where the model struggles to make the correct prediction. We will use the impact score to either add, eliminate, or use a combination of both to improve the fairness metrics.

SHAP (which stands for **Shapley Additive exPlanations**) is a model-agnostic approach in machine learning that is built on the principles of game theory. It helps study the importance of the feature and the feature interaction on the final outcome by assigning it a score, similar to how it would be done in a game where each player's contribution at a given time is calculated in the output of the score.

Shapley values can help provide global importance (the overall impact of the feature on all the predictions), but also local importance (the impact of each feature on a single outcome). It can also help understand the direction of impact – that is, whether a feature has a positive impact or a negative impact.

Hence, there are a lot of use cases for Shapley values in machine learning, such as bias detection, local and global model debugging, model auditing, and model interpretability.

We use Shapley values in this section to understand the model and feature impacts on the outcomes. We leverage the impacts of these features and identify where the model is likely to make the most mistakes. We then apply two techniques: one to remove these rows from the data and the other to oversample the data with these rows.

First, we import SHAP and then train the decision tree model on the oversampled dataset. At the end of the step, we have a model and oversampled X and y samples. We include the SEX variable in the training data to see whether Shapley values can help us detect bias. First, we need to resplit the data into train and test sets, as done in the previous sections:

```
model = DecisionTreeClassifier(**d_tree_params)
model.fit(X_train, y_train)

DecisionTreeClassifier(min_samples_leaf=10, random_state=42)
```

Next, we define the SHAP tree explainer, by providing the decision tree model and then extract the Shapley values for the train set using the .shap_values method:

```
explainer = shap.TreeExplainer(model)
shap_values = explainer.shap_values(X_train)
```

Let's extract the first row of Shapley values, for class 0. The array contains the contribution of each feature value to decide the final output. Positive values mean the corresponding features have a positive impact on predicting the output as class 0, while negative values negatively contribute toward predicting class 0:

```
shap_values[0][0]
```

This will print out the following array:

```
array([ 4.14066767e-03,  1.18731631e-02, -2.23246641e-01, -2.32653795e-02,
       -1.45181517e-02, -1.20269659e-02, -2.39415661e-04,  7.42918899e-04,
        1.03808180e-03, -6.19863882e-03, -3.95906459e-04, -3.25623889e-03,
       -2.93151055e-03, -7.50661148e-03,  6.32043125e-03, -4.79048140e-03,
       -6.07285084e-03, -8.26666488e-03,  4.60309637e-02,  9.82131153e-03,
        0.00000000e+00, -2.88275339e-05, -1.84642828e-04,  8.20571840e-05,
        0.00000000e+00,  0.00000000e+00,  0.00000000e+00,  0.00000000e+00,
        3.23840664e-04, -3.33549624e-04,  8.97779790e-06,  1.57991329e-01,
        2.96097068e-01])
```

Figure 8.13 – Resulting output array

Next, we generate a summary plot for class label 0:

```
shap.summary_plot(shap_values[0], X_train)
```

This will generate the following plot:

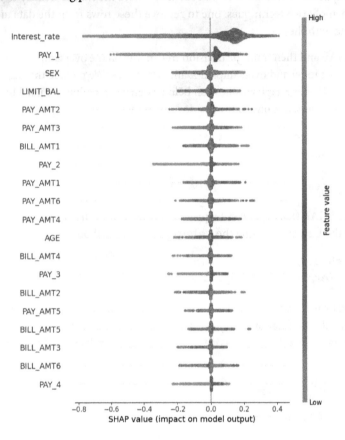

Figure 8.14 – SHAP values

The red dots represent the high value of a feature while the blue dots represent the low value of the corresponding feature. The *x* axis denotes the Shapley value, where the positive value means the data point has a positive impact in predicting class 0, whereas the negative value means the data point for the corresponding feature negatively affects the prediction for class 0. If we look at *Figure 8.13*, it is quite evident that high interest rates and male customers negatively affect the prediction of class 0. Shapley values do indicate a model bias toward male customers.

Next, we generate a summary plot for class label 1:

```
shap.summary_plot(shap_values[1], X_train)
```

This will generate the following summary plot:

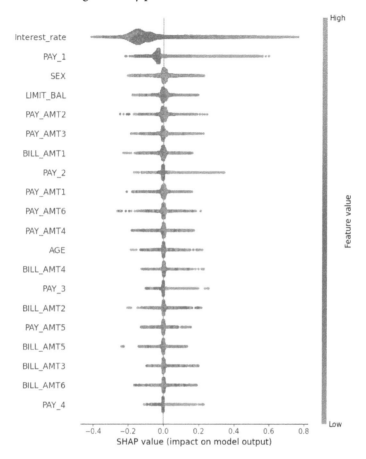

Figure 8.15 – SHAP summary plot

In comparison with the summary plot for class 0, high interest rates and male customers positively impact defaulting on the loan – that is, if you are a male and previously had a higher interest rate, you are likely to default on the loan.

We previously learned that by removing the SEX feature from model training, the model becomes fairer, and Shapley values are clearly indicated using summary plots. Now, we extract the Shapley values by training the new model without the SEX feature. We then score the training data to first identify all the rows corresponding to false negatives and false positives. We then calculate the sum of Shapley values for each row where the model made errors, and then hold out the ones with the lowest impact. We run two experiments: first, we undersample the training dataset and calculate fairness metrics, and second, we oversample the training dataset to give a better signal to the model and recalculate fairness metrics.

First, let's train the model without the SEX feature:

```
model = DecisionTreeClassifier(**d_tree_params)
X_train_samples = X_train.drop(['SEX'], axis=1).copy()
y_train_samples = y_train.copy()
model.fit(X_train_samples, y_train_samples)
```

Next, we extract Shapley values:

```
explainer = shap.Explainer(model)
shap_values = explainer.shap_values(X_train_samples)
```

We score the training data, calculate predictions, and store these in `Y_pred`:

```
Y_pred = model.predict(X_train_samples)
```

Then we will check the sum of the Shapley value for class 0 and class 1 at index 0, and print the corresponding prediction and `true` value:

```
print(f"Shapley value for first value in the dataset for class 0 :
{sum(shap_values[0][0])}")
print(f"Shapley value for first value in the dataset for class 1 :
{sum(shap_values[1][0])}")
print(f"Prediction of first value is {Y_pred[0]}")
print(f"Actual prediction is {y_train_samples[0]}")

Shapley value for first value in the dataset for class 0 :
-0.07290931372549389
Shapley value for first value in the dataset for class 1 :
0.07290931372548978
Prediction of first value is 0
Actual prediction is 1
```

The model predicted 0. Next, we extract the Shapley values where the model made a mistake. For that, we use the list comprehension with zip functionality. The first value of the array will be the index location of the data point so we know which Shapley value is associated with which row. The next values are in order of prediction, the `true` value, the sum of Shapley values for the row for class 0, and the sum of Shapley values for class 1. Once we have extracted those, we create a DataFrame and store the values in `df`, and we sample through the DataFrame to see five values. We use a random seed for reproducibility:

```
data = [(index, pred, actual, sum(s0), sum(s1)) for
        index, (pred, actual, s0, s1) in
        enumerate(zip(Y_pred, y_train_samples, shap_values[0], shap_
values[1]))
        if pred != actual]
```

```
df = pd.DataFrame(data=data, columns=["index",
"predictions","actuals", "shap_class_0", "shap_class_1"])
df.sample(5, random_state=42)
```

This generate the following output:

	index	predictions	actuals	shap_class_0	shap_class_1
422	4255	1	0	-0.478792	0.478792
1288	13255	0	1	0.087875	-0.087875
208	1972	0	1	0.110097	-0.110097
782	8169	1	0	-0.312125	0.312125
759	7915	0	1	-0.078792	0.078792

Figure 8.16 – DataFrame displaying the Shapley values that made a mistake

For index 7915, the Shapley values are close, meaning feature contributions to the model prediction are closer to 0 for each class, whereas for index 4255, the Shapley values are far apart from 0 and features are discriminatory in predicting each class.

Given that we can extract the SHAP impact of features for each class, we want to know the rows where the Shapley value impact is highest so we can eliminate such data points from training; where the impact is low and quite close to the boundary, we can oversample the data.

Looking at the force plots for index 4255, for the expected class 0, the model is likely to predict 1, given that f(x) is quite low, and the model wrongly predicts 1, whereas the force plot for the expected class 1 shows an f(x) value of 0.7. Such data points can be eliminated from the dataset:

```
shap.force_plot(explainer.expected_value[0], shap_values[0][4255,:],
X_train_samples.iloc[4255, :], matplotlib=True)
```

This will generate the following plot:

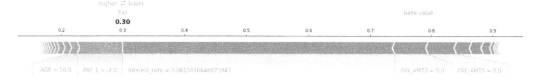

Figure 8.17 – Force plot for class 0

Let's look at the force plot for class 1:

```
shap.force_plot(explainer.expected_value[1], shap_values[1][4255,:],
X_train_samples.iloc[4255, :], matplotlib=True)
```

This will display the following output:

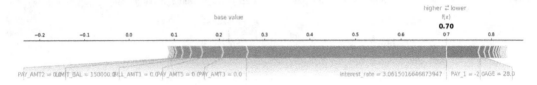

Figure 8.18 – Force plot for class 1

Now, we calculate the Shapley impact of row index `4255` since it's a false positive prediction. The row index is at the `422` location in the DataFrame. We take the absolute value of the Shapley impact and, where the Shapley impact is highest and the prediction is wrong, those values can be eliminated to improve model performance:

```
index = 422
shap_impact = abs(df['shap_class_0'][index])
print(shap_impact)
0.4787916666666662
```

Next, we create a function that calculates the Shapley impact. We are interested in those rows where a single feature has a minimum of `0.2` Shapley impact. First, we get the absolute impact of each feature in an array, and then we extract the maximum value. If the maximum value is greater than `0.2`, we proceed with that row. Next, we check where the prediction doesn't match the actual value, and for such rows, we extract the SHAP impact:

```
def get_shapley_impact(shap_value, threshold=0.2):
    """Calculate Shapley impact"""

    shap_value_impacts = np.abs(shap_value)
    if np.max(shap_value_impacts) >= threshold:
        return np.abs(np.sum(shap_value))
```

We create a holdout dataset, where `X_train` will be further divided into training and validation datasets. We leverage 80% for training and 20% for validation:

```
X_train_sample, X_val, y_train_sample, y_val, A_train_sample, A_val =
train_test_split(X_train,
                 y_train,
                 A_train,
                 test_size=0.2,
```

```
                    stratify=y_train,
                    random_state=42)
```

Then we resample the dataset using SMOTETomek, which was the best sampling method for fairness and performance, and performed by adding difficult examples back to the dataset. Once the dataset is resampled, we train the standard decision tree as in the previous steps, and calculate the ROC score on the holdout validation dataset:

```
model = DecisionTreeClassifier(**d_tree_params)

scaler = StandardScaler()
sampler = SMOTETomek(random_state=42,
                     smote=SMOTE(random_state=42),
                     tomek=TomekLinks(n_jobs=-1)
                     )

columns = X_train_sample.columns

X_train_scaled = pd.DataFrame()
X_train_scaled[columns] = scaler.fit_transform(X_train_
sample[columns])

X_train_resampled, y_train_resampled = sampler_method.fit_resample(X_
train_scaled, y_train_sample)
X_train_resampled[columns] = scaler.inverse_transform(X_train_
resampled)
A_train_resampled = X_train_resampled['SEX'].copy()

model.fit(X_train_resampled.drop(['SEX'], axis=1), y_train_resampled)
Y_pred = model.predict(X_train_resampled.drop(['SEX'], axis=1))
y_val_pred = model.predict_proba(X_val.drop(['SEX'], axis=1))[:,1]
val_roc = roc_auc_score(y_val, y_val_pred)

print(f"Validation roc auc : {val_roc}")
Validation roc auc : 0.8275769341994823
```

We calculate the fairness and performance metrics on the testing dataset. The equalized odds are high but the demographic parity ratio is within the accepted range:

```
y_pred_proba = model.predict_proba(X_test.drop(['SEX'], axis=1))[:, 1]
y_pred = model.predict(X_test.drop(['SEX'], axis=1))
print(f"Roc: {roc_auc_score(y_test, y_pred_proba)}")

Roc: 0.8258396009334518
```

Next, we calculate the fairness metrics:

```
calculate_fairness_metrics(y_test, y_pred, A_test)
```

This will print out the following metrics:

```
            balanced_accuracy  false_positive_rate  false_negative_rate
SEX
female             0.788349             0.146540             0.276762
male               0.694428             0.224335             0.386809

*diff*
balanced_accuracy    0.093921
false_positive_rate  0.077795
false_negative_rate  0.110047
dtype: float64

*final_metrics*
balanced_accuracy    0.749763
false_positive_rate  0.177188
false_negative_rate  0.323286
dtype: float64

*equalized_odds*
0.11004686707343747

*demographic_parity_ratio*
0.8545544530463497
```

Figure 8.19 – Fairness metrics

We extract the Shapley values using the SHAP explainer. We have ensured that the SEX feature is removed:

```
explainer = shap.Explainer(model)
columns = X_train_resampled.drop(['SEX'], axis=1).columns
shap_values = explainer.shap_values(X_train_resampled[columns])
```

The fairness metrics demonstrate that the gap between false negative rates is higher between the subclass of men and women, hence we focus on reducing false negative cases for males using Shapley values.

In the next step, we extract Shapley values where the model predicts class 0, but the true value is 1. Hence, we are interested in Shapley values for class 1, as a high SHAP impact for class 1 where the model made an error could be a data point that the model is unable to make a correct prediction on:

```
shapley_impact_false_negative = [(i, get_shapley_impact(s), y, p, a)
                          for s, i, y, p, a
                          in zip(shap_values[1], X_train_
resampled.index, y_train_resampled, Y_pred, A_train_resampled)
                          if y == 1 and p == 0 and a == 1]
```

We are also interested in those data points where both Shapley values and model prediction agree with the actual values. Hence, we focus on those data points where the model rightly predicts class

0 for `male` data points. Once we have extracted those, we focus on high-impact Shapley values for class 0 so we can oversample the dataset with those, such that the model can get a better signal for such data points:

```
shapley_impact_true_negative = [(i, get_shapley_impact(s), y, p, a)
                                for s, i, y, p, a
                                in zip(shap_values[0], X_train_
resampled.index, y_train_resampled, Y_pred, A_train_resampled)
                                if y == 0 and p == 0 and a == 1]
```

Next, we sort the false negative Shapley values so we can extract the high-impact data points:

```
shapley_impact_false_negative_sorted = sorted([i for i in shapley_
impact_false_negative if i[1] is not None], key=lambda x: x[1],
reverse=True)
```

Similar to the preceding, we are interested in true negative high-impact Shapley values, so we sort the list according to high-impact Shapley values:

```
shapley_impact_true_negative_sorted =  sorted([i for i in shapley_
impact_true_negative if i[1] is not None], key=lambda x: x[1],
reverse=True)
```

Now that we have extracted and sorted the Shapley values for false negative and true negative `male` data points, we pick the top 100 data points to eliminate from the false negative list and pick the top 100 data points from the true negative list to add back to the training data. The top 100 data points from the true negative list will be shuffled and only 50 data points from there will be added at random with a replacement strategy. We encourage practitioners to try another ratio for shuffling. Once the data points are identified for elimination and reintroduction to the final training set, we update the training data. These are named `X_train_final` and `y_train_final`:

```
data_points_to_eliminate = [i[0] for i in shapley_impact_false_
negative_sorted[0:100]]
data_points_to_add = [i[0] for i in shapley_impact_true_negative_
sorted[0:100]]
X_train_added = X_train_resampled[columns].iloc[data_points_to_add].
sample(frac=0.5, replace=True, random_state=42, axis=0)
y_train_added = y_train_resampled.iloc[data_points_to_add].
sample(frac=0.5, replace=True, random_state=42, axis=0)

X_train_reduced = X_train_resampled[columns].drop(data_points_to_
eliminate)
y_train_reduced = y_train_resampled.drop(data_points_to_eliminate)

X_train_final = pd.concat([X_train_reduced, X_train_added], axis=0,
ignore_index=True)
y_train_final = pd.concat([y_train_reduced, y_train_added], axis=0,
ignore_index=True)
```

Next, we train the updated training data and calculate fairness metrics and performance metrics on the test data. It is evident that the gap between false negative rates has reduced, the equalized odds ratio has improved to 0.082, and the ROC score has slightly improved from the previous step, from 0.825 to 0.826:

```
estimator = DecisionTreeClassifier(**d_tree_params)
model = estimator.fit(X_train_final, y_train_final)

y_pred_proba = model.predict_proba(X_test.drop(['SEX'], axis=1))[:, 1]
y_pred = model.predict(X_test.drop(['SEX'], axis=1))
print(f"Roc: {roc_auc_score(y_test, y_pred_proba)}")

Roc: 0.8262453372973797
```

We recalculate the fairness metrics:

```
calculate_fairness_metrics(y_test, y_pred, A_test)
```

The output is as follows:

```
            balanced_accuracy  false_positive_rate  false_negative_rate
SEX
female               0.785845             0.143715             0.284595
male                 0.706405             0.219989             0.367201

 *diff*
balanced_accuracy      0.079440
false_positive_rate    0.076274
false_negative_rate    0.082606
dtype: float64

 *final_metrics*
balanced_accuracy      0.753359
false_positive_rate    0.173764
false_negative_rate    0.319518
dtype: float64

 *equalized_odds*
0.08260612576385884

 *demographic_parity_ratio*
0.8388834440186068
```

Figure 8.20 – Fairness metrics

Now that we have determined that by using Shapley values we can identify data points that are difficult to classify and easy to classify, we can build an automatic mechanism to iterate over the data points such that we can reach a better fairness score than previously.

Next, we create a range of percentages to iterate so we can leverage and sort through the top data points as a percentage of top data points to eliminate and reintroduce. We will leverage the NumPy `linspace` method to create a list of percentage values to iterate. We choose 10 values from `0.05` through `0.5` (5 to 50 percent). We call this list `perc_points_to_eliminate`:

```
perc_points_to_eliminate = np.linspace(0.05,0.5,10)
perc_points_to_eliminate
```

```
array([0.05, 0.1 , 0.15, 0.2 , 0.25, 0.3 , 0.35, 0.4 , 0.45, 0.5 ])
```

We iterate through these percentages and repeat the preceding step where we eliminated some values and reintroduced some values. However, this time, instead of 100, we use percentages to remove the top percent of data points or introduce the top percent of data points.

We also create an empty list of data so, for each iteration, we capture the percentage of data points eliminated or reintroduced, false negative and false positive rates on test data, equalized odds ratio, and demographic parity ratio.

Once we have iterated over all the values – 10*10 iterations, we store them in a DataFrame to see how many data points need to be removed and how many added to lead to the best fairness metrics:

```
fn_examples = len(shapley_impact_false_negative)
tn_examples = len(shapley_impact_true_negative)
model = DecisionTreeClassifier(**d_tree_params)

data = []
for fnp in perc_points_to_eliminate:
    data_points_to_eliminate = [idx[0] for idx in shapley_impact_
false_negative_sorted[0:(round(fn_examples*fnp))]]
    for tnp in perc_points_to_eliminate:
        data_points_to_add = [idx[0] for idx in shapley_impact_true_
negative_sorted[0:(round(tn_examples*tnp))]]

        X_train_added = X_train_resampled[columns].iloc[data_points_
to_add].sample(frac=0.5, replace=True, random_state=42, axis=0)
        y_train_added = y_train_resampled.iloc[data_points_to_add].
sample(frac=0.5, replace=True, random_state=42, axis=0)

        X_train_reduced = X_train_resampled[columns].drop(data_points_
to_eliminate)
        y_train_reduced = y_train_resampled.drop(data_points_to_
eliminate)
```

```
        X_train_final = pd.concat([X_train_reduced, X_train_added],
axis=0, ignore_index=True)
        y_train_final = pd.concat([y_train_reduced, y_train_added],
axis=0, ignore_index=True)

        model.fit(X_train_final, y_train_final)

        y_pred = model.predict(X_test.drop(['SEX'], axis=1))
        fpr = false_positive_rate(y_test, y_pred)
        fnr = false_negative_rate(y_test, y_pred)

        equalized_odds_mitigated = equalized_odds_difference(
            y_test, y_pred, sensitive_features=A_test
        )

        demographic_parity_ratio_mitigated = demographic_parity_
ratio(y_test, y_pred, sensitive_features=A_test)
        data.append((fnp,
                    tnp,
                    fpr,
                    fnr,
                    equalized_odds_mitigated,
                    demographic_parity_ratio_mitigated
                ))
```

Next, we create a DataFrame called `df_shapley` for the metadata for each iteration and sort it by equalized odds ratio:

```
columns = ["perc_false_negative_removed",
           "perc_true_negative_added",
           "false_positive_rate",
           "false_negative_rate",
           "equalized_odds_mitigated",
           "demographic_parity_ratio_mitigated"]

df_shapley = pd.DataFrame(data=data, columns=columns)
df_shapley.sort_values(by="equalized_odds_mitigated")
```

This will output the following DataFrame:

	perc_false_negative_removed	perc_true_negative_added	false_positive_rate	false_negative_rate	equalized_odds_mitigated	demographic_parity_ratio_mitigat
45	0.25	0.30	0.165846	0.343632	0.074799	0.854:
44	0.25	0.25	0.166916	0.343632	0.077887	0.8505
34	0.20	0.25	0.165846	0.339864	0.081326	0.8484
79	0.40	0.50	0.160068	0.332329	0.082029	0.851€
25	0.15	0.30	0.170126	0.335343	0.082983	0.8473
...	...	...	...	...	...	
15	0.10	0.30	0.164562	0.335343	0.113863	0.861
70	0.40	0.05	0.170768	0.317257	0.114315	0.8532
61	0.35	0.10	0.167130	0.314996	0.115143	0.8427
67	0.35	0.40	0.165204	0.326300	0.120265	0.851€
60	0.35	0.05	0.166488	0.309721	0.121194	0.8422

100 rows × 6 columns

Figure 8.21 – The df_shapley DataFrame after sorting

It's evident that when the top 25% of false negative data points are removed and 30% of the top true negative data points are reintroduced, the model can achieve an equalized odds ratio of 0.074 with an optimum demographic parity ratio score of 0.85.

Finally, we extract the top percentages and train the final model:

```
top_values = df_shapley.sort_values(by="equalized_odds_mitigated").
values[0]
perc_false_negative_removed = top_values[0]
perc_true_negative_added = top_values[1]


columns = X_train_resampled.drop(['SEX'], axis=1).columns
data_points_to_eliminate = [i[0] for i in shapley_impact_false_
negative_sorted[0:(round(fn_examples*perc_false_negative_removed))]]
data_points_to_add = [i[0] for i in shapley_impact_true_negative_
sorted[0:(round(tn_examples*perc_true_negative_added))]]
X_train_added = X_train_resampled[columns].iloc[data_points_to_add].
sample(frac=0.5, replace=True, random_state=42, axis=0)
y_train_added = y_train_resampled.iloc[data_points_to_add].
sample(frac=0.5, replace=True, random_state=42, axis=0)


X_train_reduced = X_train_resampled[columns].drop(data_points_to_
eliminate)
y_train_reduced = y_train_resampled.drop(data_points_to_eliminate)


X_train_final = pd.concat([X_train_reduced, X_train_added], axis=0,
ignore_index=True)
y_train_final = pd.concat([y_train_reduced, y_train_added], axis=0,
ignore_index=True)
```

We train the model and calculate the fairness and model metrics. We can see that the false negative rate for `female` has increased but the gap between `male` and `female` has reduced, and the false positive rate for `male` has reduced. The ROC score achieved is 0.82, but the model is much fairer based on two fairness metrics:

```
estimator = DecisionTreeClassifier(**d_tree_params)
model = estimator.fit(X_train_final, y_train_final)

y_pred_proba = model.predict_proba(X_test.drop(['SEX'], axis=1))[:, 1]
y_pred = model.predict(X_test.drop(['SEX'], axis=1))
print(f"Roc: {roc_auc_score(y_test, y_pred_proba)}")

Roc: 0.820911097453972
```

Finally, we calculate the fairness metrics.

```
calculate_fairness_metrics(y_test, y_pred, A_test)
```

This will print out the following:

```
            balanced_accuracy  false_positive_rate  false_negative_rate
SEX
female            0.773903           0.140184             0.312010
male              0.703934           0.205323             0.386809

 *diff*
balanced_accuracy     0.069969
false_positive_rate   0.065140
false_negative_rate   0.074799
dtype: float64

 *final_metrics*
balanced_accuracy     0.745261
false_positive_rate   0.165846
false_negative_rate   0.343632
dtype: float64

 *equalized_odds*
0.07479882529798065

 *demographic_parity_ratio*
0.8543712342498679
```

Figure 8.22 – Fairness metrics

Now that we have explored different data-centric techniques for reducing bias by improving data quality, we encourage you to experiment with the previous techniques and try a combination of these. Once you have exhausted these data-centric approaches, we encourage you use some model-centric approaches, such as utilizing algorithms that are fairness aware and trying ensembling methods, AutoML, or iterating through your own list of algorithms.

Summary

This chapter provided an extensive exploration of the pervasive challenge of bias in machine learning. It started by explaining various forms of bias inherent in machine learning models and examined their impact on different industries. The emphasis was on recognizing, monitoring, and mitigating bias, underscoring the importance of collecting data with minimal selection and sampling bias.

The central theme advocated a data-centric imperative over a model-centric one in addressing bias. Techniques such as oversampling, undersampling, feature selection enhancement, and anomaly detection were explored for bias rectification. Shapley values play a crucial role in bias identification, emphasizing the removal of examples with misaligned high Shapley values and the reintroduction of data points with replacement to improve ratios. Stratification of misclassified examples based on sensitive variables such as SEX was outlined for targeted bias correction.

The chapter concluded by highlighting the significance of refining and balancing datasets concerning sensitive variables as a foundational step. It suggested progressing toward model-centric approaches, such as ensembling and fairness algorithms, once the dataset itself has been improved. These subsequent model-centric strategies aim to enhance both performance and fairness metrics, establishing a foundation for more generalized and equitable AI models.

This comprehensive approach strives to create a balanced dataset as a precursor to applying model-centric techniques, promoting performance and fairness in AI systems.

9

Dealing with Edge Cases and Rare Events in Machine Learning

In the field of **machine learning**, it is crucial to identify and handle edge cases properly. Edge cases refer to instances in your dataset that are significantly different from the majority of the data, and they can have a substantial impact on the performance and reliability of your machine learning models. Rare events can be challenging for machine learning models due to class imbalance problems as they might not have enough data to learn patterns effectively. Class imbalance occurs when one class (the rare event) is significantly underrepresented compared to the other class(es). Traditional machine learning algorithms tend to perform poorly in such scenarios because they may be biased toward the majority class, leading to lower accuracy in identifying rare events.

In this chapter, we will explore various techniques and approaches to detect edge cases in machine learning and data, using Python code examples. We'll explore statistical techniques, including using visualizations and other measures such as Z-scores, to analyze data distributions and identify potential outliers. We will also focus on methodologies such as isolation forests and semi-supervised methods such as **autoencoders** to uncover anomalies and irregular patterns in your datasets. We will learn how to address class imbalance and enhance model performance through techniques such as oversampling, undersampling, and generating synthetic data.

In the latter half of this chapter, we'll understand the importance of adjusting the learning process to account for imbalanced classes, especially in scenarios where the rare event carries significant consequences. We'll explore the significance of selecting appropriate evaluation metrics, emphasizing those that account for class imbalance, to ensure a fair and accurate assessment of model performance. Finally, we'll understand how ensemble methods such as bagging, boosting, and stacking help us enhance the robustness of models, particularly in scenarios where rare events play a crucial role.

The following key topics will be covered:

- Importance of detecting rare events and edge cases in machine learning
- Statistical methods
- Anomaly detection
- Data augmentation and resampling techniques
- Cost-sensitive learning
- Choosing evaluation metrics
- Ensemble techniques

Importance of detecting rare events and edge cases in machine learning

Detecting rare events and edge cases is crucial in machine learning for several reasons:

- **Decision-making in critical scenarios**: Rare events often represent critical scenarios or anomalies that require immediate attention or special treatment. For instance, in medical diagnosis, rare diseases or extreme cases might need urgent intervention. Accurate detection of these events can lead to better decision-making and prevent adverse consequences.

- **Unbalanced datasets**: Many real-world datasets suffer from class imbalance, where one class (often the rare event) is significantly underrepresented compared to the other classes. This can lead to biased models that perform poorly on the minority class. Detecting rare events helps identify the need for special handling, such as using resampling techniques or employing appropriate evaluation metrics to ensure fair evaluation.

- **Fraud detection**: In fraud detection applications, rare events often correspond to fraudulent transactions or activities. Detecting these rare cases is crucial for preventing financial losses and ensuring the security of financial systems.

- **Quality control and anomaly detection**: In manufacturing and industrial processes, detecting rare events can help identify faulty or anomalous products or processes. This enables timely intervention to improve product quality and maintain operational efficiency.

- **Predictive maintenance**: In predictive maintenance, detecting edge cases can indicate potential equipment failures or abnormal behaviors, allowing proactive maintenance to reduce downtime and increase productivity.

- **Model generalization**: By accurately identifying and handling rare events, machine learning models can better generalize to unseen data and handle real-world scenarios effectively.

- **Customer behavior analysis**: In marketing and customer analysis, detecting rare events can reveal unusual patterns or behaviors of interest, such as identifying high-value customers or detecting potential churners.

- **Security and intrusion detection**: In cybersecurity, rare events may indicate security breaches or cyber-attacks. Detecting and responding to these events in real time is essential for ensuring the safety and integrity of digital systems.

- **Environmental monitoring**: In environmental applications, rare events might signify unusual ecological conditions or natural disasters. Detecting such events aids in disaster preparedness and environmental monitoring efforts.

Let's discuss different statistical methods to analyze data distributions and identify potential outliers.

Statistical methods

Statistical methods provide valuable tools for identifying outliers and anomalies in our data, aiding in data preprocessing and decision-making. In this section, we'll talk about how to use methods such as Z-scores, **Interquartile Range (IQR)**, box plots, and scatter plots to uncover anomalies in our data.

Z-scores

Z-scores, also known as standard scores, are a statistical measure that indicates how many standard deviations a data point is away from the mean of the data. Z-scores are used to standardize data and allow for comparisons between different datasets, even if they have different units or scales. They are particularly useful in detecting outliers and identifying extreme values in a dataset. The formula to calculate the Z-score for a data point x in a dataset with mean μ and standard deviation σ is presented here:

$$Z = (x - \mu)/\sigma$$

Here, the following applies:

- Z is the Z-score of the data point x

- x is the value of the data point

- μ is the mean of the dataset

- σ is the standard deviation of the dataset

Z-scores are widely used to detect outliers in a dataset. Data points with Z-scores that fall outside a certain threshold (e.g., $Z > 3$ or $Z < -3$) are considered outliers. These outliers can represent extreme values or measurement errors in the data. By transforming the data into Z-scores, the mean becomes 0, and the standard deviation becomes 1, resulting in a standardized distribution.

Z-scores are employed in normality testing to assess whether a dataset follows a normal (Gaussian) distribution. If the dataset follows a normal distribution, approximately 68% of the data points should have Z-scores between -1 and 1, about 95% between -2 and 2, and nearly all between -3 and 3. In hypothesis testing, Z-scores are used to compute *p*-values and make inferences about population parameters.

For example, Z-tests are commonly used for sample mean comparisons when the population standard deviation is known. Z-scores can be useful in anomaly detection where we want to identify data points that deviate significantly from the norm. High Z-scores may indicate anomalous behavior or rare events in the data.

Let's explore this concept in Python using the *Loan Prediction* dataset. Let's start by loading the dataset:

```
import pandas as pd
import numpy as np
df = pd.read_csv('train_loan_prediction.csv')
df.head().T
```

Here, we see the sample dataset:

	0	1	2	3	4
Loan_ID	LP001002	LP001003	LP001005	LP001006	LP001008
Gender	Male	Male	Male	Male	Male
Married	No	Yes	Yes	Yes	No
Dependents	0	1	0	0	0
Education	Graduate	Graduate	Graduate	Not Graduate	Graduate
Self_Employed	No	No	Yes	No	No
ApplicantIncome	5849	4583	3000	2583	6000
CoapplicantIncome	0.0	1508.0	0.0	2358.0	0.0
LoanAmount	NaN	128.0	66.0	120.0	141.0
Loan_Amount_Term	360.0	360.0	360.0	360.0	360.0
Credit_History	1.0	1.0	1.0	1.0	1.0
Property_Area	Urban	Rural	Urban	Urban	Urban
Loan_Status	Y	N	Y	Y	Y

Figure 9.1 – The df DataFrame

Now that we have loaded the dataset, we can calculate Z-scores on some of the numerical features:

```
numerical_features = ['ApplicantIncome', 'CoapplicantIncome',
'LoanAmount']
z_scores = df[numerical_features].apply(lambda x: (x - np.mean(x)) /
np.std(x))
```

We leverage Z-scores to detect outliers in our dataset effectively:

```
threshold = 3
outliers = (z_scores > threshold) | (z_scores < -threshold)
outliers['is_outlier'] = outliers.any(axis=1)
outlier_rows = df[outliers['is_outlier']]
outlier_rows
```

By setting a Z-score threshold of 3, we can identify outliers by examining if any rows have values that lie beyond the Z-score boundaries of greater than 3 or less than -3. This approach allows us to pinpoint data points that deviate significantly from the mean and enables us to take appropriate actions to handle these outliers, ensuring the integrity of our data and the accuracy of subsequent analyses and models.

Here is the output DataFrame of outliers using this approach:

	Loan_ID	Gender	Married	Dependents	Education	Self_Employed	ApplicantIncome	CoapplicantIncome	LoanAmount	Loan_Amount_Term	Credit_Hi
9	LP001020	Male	Yes	1	Graduate	No	12841	10968.0	349.0	360.0	1.00
126	LP001448	NaN	Yes	3+	Graduate	No	23803	0.0	370.0	360.0	1.00
130	LP001469	Male	No	0	Graduate	Yes	20166	0.0	650.0	480.0	0.84
155	LP001536	Male	Yes	3+	Graduate	No	39999	0.0	600.0	180.0	0.00
171	LP001585	NaN	Yes	3+	Graduate	No	51763	0.0	700.0	300.0	1.00
177	LP001610	Male	Yes	3+	Graduate	No	5516	11300.0	495.0	360.0	0.00
183	LP001637	Male	Yes	1	Graduate	No	33846	0.0	260.0	360.0	1.00
185	LP001640	Male	Yes	0	Graduate	Yes	39147	4750.0	120.0	360.0	1.00
278	LP001907	Male	Yes	0	Graduate	No	14583	0.0	436.0	360.0	1.00
308	LP001996	Male	No	0	Graduate	No	20233	0.0	480.0	360.0	1.00
333	LP002101	Male	Yes	0	Graduate	NaN	63337	0.0	490.0	180.0	1.00
369	LP002191	Male	Yes	0	Graduate	No	19730	5266.0	570.0	360.0	1.00
402	LP002297	Male	No	0	Graduate	No	2500	20000.0	103.0	360.0	1.00
409	LP002317	Male	Yes	3+	Graduate	No	81000	0.0	360.0	360.0	0.00
417	LP002342	Male	Yes	2	Graduate	Yes	1600	20000.0	239.0	360.0	1.00
432	LP002386	Male	No	0	Graduate	NaN	12876	0.0	405.0	360.0	1.00
443	LP002422	Male	No	1	Graduate	No	37719	0.0	152.0	360.0	1.00
487	LP002547	Male	Yes	1	Graduate	No	18333	0.0	500.0	360.0	1.00
506	LP002624	Male	Yes	0	Graduate	No	20833	6667.0	480.0	360.0	0.84
523	LP002693	Male	Yes	2	Graduate	Yes	7948	7166.0	480.0	360.0	1.00
525	LP002699	Male	Yes	2	Graduate	Yes	17500	0.0	400.0	360.0	1.00
561	LP002813	Female	Yes	1	Graduate	Yes	19484	0.0	600.0	360.0	1.00
581	LP002893	Male	No	0	Graduate	No	1836	33837.0	90.0	360.0	1.00
600	LP002949	Female	No	3+	Graduate	NaN	416	41667.0	350.0	180.0	0.84
604	LP002959	Female	Yes	1	Graduate	No	12000	0.0	496.0	360.0	1.00

Figure 9.2 – The resulting outlier_rows DataFrame

In conclusion, Z-scores make it easier to spot outliers accurately. They act as a helpful tool, guiding us to understand data patterns better. In the next section on IQR, we will further explore alternative methods for detecting and addressing outliers.

Interquartile Range (IQR)

IQR is a statistical measure used to describe the spread or dispersion of a dataset. It is particularly useful in identifying and handling outliers and understanding the central tendency of the data. The IQR is defined as the range between the first quartile (Q1) and the third quartile (Q3) of a dataset. Quartiles are points that divide a dataset into four equal parts, each containing 25% of the data.

The IQR can be calculated using the following formula:

$$IQR = Q3 - Q1$$

Here, the following applies:

- Q1 is the first quartile (25th percentile), representing the value below which 25% of the data lies

- Q3 is the third quartile (75th percentile), representing the value below which 75% of the data lies

IQR is commonly used to identify outliers in a dataset. Data points that fall below Q1 - 1.5 * IQR or above Q3 + 1.5 * IQR are considered outliers and may warrant further investigation. IQR provides valuable information about the distribution of the data. It helps to understand the spread of the middle 50% of the dataset and can be used to assess the symmetry of the distribution. When comparing datasets, IQR can be used to assess differences in the spread of data between two or more datasets.

Here's a simple example of how to calculate the IQR for a dataset using Python and NumPy on the Loan Prediction dataset. To begin with outlier detection, we calculate the IQR for numerical features. By defining a threshold, typically 1.5 times the IQR based on Tukey's method, which involves calculating the IQR as the difference between the third quartile (Q3) and the first quartile (Q1), we can identify outliers. Tukey's method is a robust statistical technique that aids in detecting data points that deviate significantly from the overall distribution, providing a reliable measure for identifying potential anomalies in the dataset. Once outliers are flagged for all numerical features, we can display the rows with outlier values, which aids in further analysis and potential data treatment. Using the IQR and Tukey's method together facilitates effective outlier detection and helps ensure the integrity of the dataset:

```
Q1 = df[numerical_features].quantile(0.25)
Q3 = df[numerical_features].quantile(0.75)
IQR = Q3 - Q1
threshold = 1.5
outliers = (df[numerical_features] < (Q1 - threshold * IQR)) |
(df[numerical_features] > (Q3 + threshold * IQR))
outliers['is_outlier'] = outliers.any(axis=1)
outlier_rows = df[outliers['is_outlier']]
outlier_rows
```

Here is the output DataFrame of outliers using this approach:

	Loan_ID	Gender	Married	Dependents	Education	Self_Employed	ApplicantIncome	CoapplicantIncome	LoanAmount	Loan_Amount_Term	(
9	LP001020	Male	Yes	1	Graduate	No	12841	10968.0	349.0	360.0	
12	LP001028	Male	Yes	2	Graduate	No	3073	8106.0	200.0	360.0	
21	LP001046	Male	Yes	1	Graduate	No	5955	5625.0	315.0	360.0	
34	LP001100	Male	No	3+	Graduate	No	12500	3000.0	320.0	360.0	
38	LP001114	Male	No	0	Graduate	No	4166	7210.0	184.0	360.0	
...	...	...	...	...	...	...	...	...	...	...	
581	LP002893	Male	No	0	Graduate	No	1836	33837.0	90.0	360.0	
592	LP002933	NaN	No	3+	Graduate	Yes	9357	0.0	292.0	360.0	
594	LP002938	Male	Yes	0	Graduate	Yes	16120	0.0	260.0	360.0	
600	LP002949	Female	No	3+	Graduate	NaN	416	41667.0	350.0	180.0	
604	LP002959	Female	Yes	1	Graduate	No	12000	0.0	496.0	360.0	

Figure 9.3 – The resulting outlier_rows DataFrame

In conclusion, the exploration of IQR provides a valuable technique for identifying outliers, particularly effective in capturing the central tendencies of a dataset. While IQR proves advantageous in scenarios where a focus on the middle range is paramount, it's essential to recognize its limitations in capturing the entire data distribution. In the upcoming section on box plots, we will delve into a graphical representation that complements IQR, offering a visual tool to better understand data dispersion and outliers in diverse contexts.

Box plots

Box plots, also known as box-and-whisker plots, are a graphical representation of the distribution of a dataset. They provide a quick and informative way to visualize the spread and skewness of the data, identify potential outliers, and compare multiple datasets. Box plots are particularly useful when dealing with continuous numerical data and can be used to gain insights into the central tendency and variability of the data.

Here are the components of a box plot:

- **Box (IQR)**: The box represents the IQR, which is the range between the first quartile (Q1) and the third quartile (Q3) of the data. It spans the middle 50% of the dataset and provides a visual representation of the data's spread.

- **Median (Q2)**: The median, represented by a horizontal line inside the box, indicates the central value of the dataset. It divides the data into two equal halves, with 50% of the data points below and 50% above the median.

- **Whiskers**: Whiskers extend from the edges of the box to the furthest data points that lie within the "whisker length." The length of the whiskers is typically determined by a factor (for example, 1.5 times the IQR) and is used to identify potential outliers.

- **Outliers**: Data points lying beyond whiskers are considered outliers and are usually plotted individually as individual points or circles. They are data points that deviate significantly from the central distribution and may warrant further investigation.

The following are the benefits of box plots:

- **Visualizing data distribution**: Box plots offer an intuitive way to see the spread and skewness of the data, as well as identify any potential data clusters or gaps
- **Comparing datasets**: Box plots are useful for comparing multiple datasets side by side, allowing for easy comparisons of central tendencies and variabilities
- **Outlier detection**: Box plots facilitate outlier detection by highlighting data points that lie beyond whiskers, helping identify unusual or extreme values
- **Handling skewed data**: Box plots are robust to the influence of extreme values and can handle skewed data distributions more effectively than traditional mean and standard deviation

Let's implement this approach using the `matplotlib` library in Python. Here, we will produce two box plots for `ApplicantIncome` and `LoanAmount`:

```
import matplotlib.pyplot as plt
%matplotlib inline
plt.figure(figsize=(10, 6))
plt.subplot(2, 1, 1)
df.boxplot(column='ApplicantIncome')
plt.title('Box Plot - ApplicantIncome')
plt.subplot(2, 1, 2)
df.boxplot(column='LoanAmount')
plt.title('Box Plot - LoanAmount')

plt.tight_layout()
plt.show()
```

Here are the plot outputs. We see that there are some possible outliers in these columns:

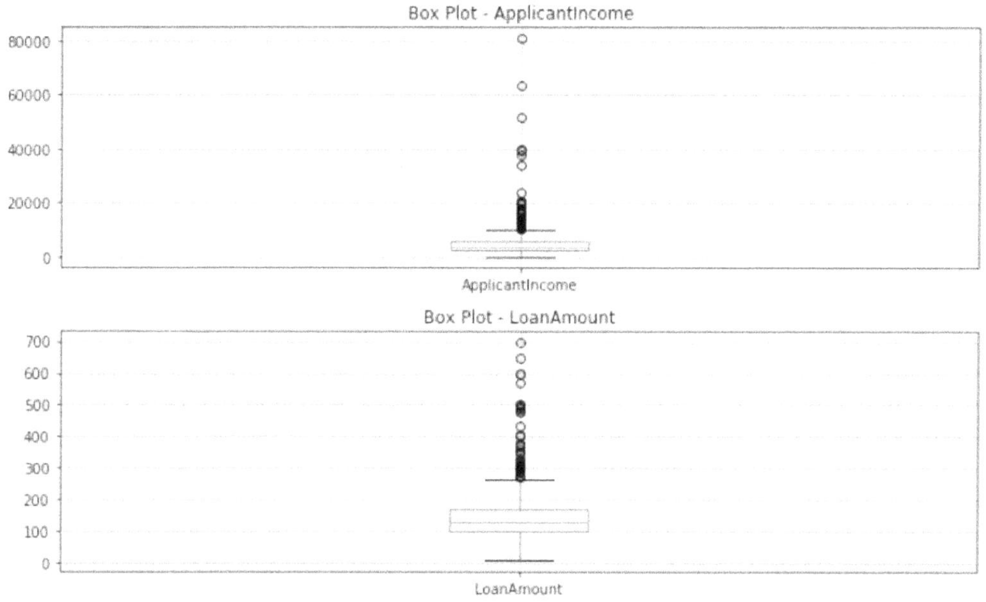

Figure 9.4 – Box plots for ApplicantIncome (top) and LoanAmount (bottom)

In summarizing the use of box plots, we find them to be a powerful visual aid, complementing IQR in portraying both central tendencies and data dispersion. While box plots excel in providing a holistic view, it's crucial to acknowledge that they may not capture all nuances of complex datasets. In the next section on scatter plots, we will explore a versatile graphical tool that offers a broader perspective, facilitating the identification of relationships and patterns between variables in our data.

Scatter plots

Scatter plots are a popular and versatile data visualization technique used to explore the relationship between two continuous numerical variables. They provide a clear visual representation of how one variable (the independent variable) affects or influences another (the dependent variable). Scatter plots are especially useful in identifying patterns, correlations, clusters, and outliers in data, making them an essential tool in data analysis and **exploratory data analysis** (**EDA**).

Let's now look at how scatter plots are constructed and their key characteristics.

Scatter plot construction

To create a scatter plot, the values of two numerical variables are plotted as points on a Cartesian coordinate system. Each point represents a data observation, where the x coordinate corresponds to the value of the independent variable and the y coordinate corresponds to the value of the dependent variable. Multiple data points collectively form a scatter plot that provides insights into the relationship between the two variables.

Key characteristics of scatter plots

The following are some key characteristics of scatter plots:

- **Correlation**: Scatter plots help us assess the correlation or relationship between two variables. If points on the plot appear to form a clear trend or pattern (e.g., a linear or non-linear trend), it suggests a significant correlation between the variables. If points are scattered randomly, there might be no or weak correlation.

- **Cluster analysis**: Scatter plots can reveal clusters of data points, indicating potential subgroups or patterns within the data.

- **Outlier detection**: Scatter plots facilitate outlier detection by identifying data points that lie far away from the main cluster of points.

- **Data spread**: The spread or distribution of data points along the x and y axes provides insights into the variability of the variables.

- **Visualizing regression lines**: In some cases, a regression line can be fitted to the scatter plot to model the relationship between the variables and make predictions.

Let's implement this in Python on the `ApplicantIncome` and `LoanAmount` columns:

```
plt.figure(figsize=(8, 6))
plt.scatter(df['ApplicantIncome'], df['LoanAmount'])
plt.xlabel('ApplicantIncome')
plt.ylabel('LoanAmount')
plt.title('Scatter Plot - ApplicantIncome vs. LoanAmount')
plt.show()
```

Here is the output showcasing the scatter plot results:

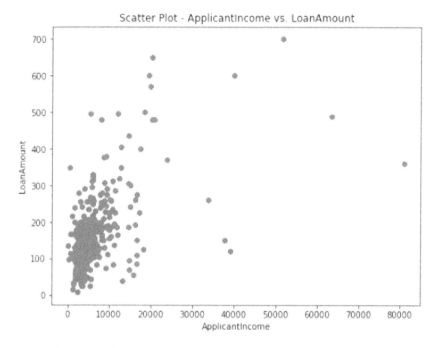

Figure 9.5 – Scatter plot showcasing ApplicantIncome and LoanAmount

As you can see, some points are notably distant from the majority of the population, suggesting the presence of potential outliers. These outlying data points stand apart from the overall pattern, warranting further investigation to understand their significance and impact on the relationship between the two variables. Identifying and handling outliers is crucial for ensuring accurate data analysis and model performance. By visualizing the scatter plot, we can gain valuable insights into data distribution and correlations, paving the way for effective decision-making and data exploration.

Anomaly detection

Anomaly detection is a specific approach to detecting rare events, where the focus is on identifying instances that significantly deviate from the norm or normal behavior. Anomalies can be caused by rare events, errors, or unusual patterns that are not typical in the dataset. This technique is particularly useful when there is limited or no labeled data for rare events. Common anomaly detection algorithms include the following:

- **Unsupervised methods**: Techniques such as Isolation Forest and **One-Class SVM**) can be used to identify anomalies in data without requiring labeled examples of the rare event.

- **Semi-supervised methods**: These approaches combine normal and abnormal data during training but have only a limited number of labeled anomalies. Autoencoders and variational autoencoders are examples of semi-supervised anomaly detection algorithms.

- **Supervised methods**: If a small number of labeled anomalies are available, **supervised learning** algorithms such as Random Forest, **Support Vector Machines** (**SVM**), and neural networks can be used for anomaly detection.

Let's understand these methods in detail with Python code examples.

Unsupervised method using Isolation Forest

Isolation Forest is an efficient and effective algorithm used for anomaly detection in **unsupervised learning** scenarios. It works by isolating anomalies or rare events in the data by constructing isolation trees (random decision trees) that separate the anomalies from the majority of the normal data points. It was introduced by Fei Tony Liu, Kai Ming Ting, and Zhi-Hua Zhou in their 2008 paper titled *Isolation Forest*. Here are some key concepts and features of the Isolation Forest algorithm:

- **Random partitioning**: Isolation Forest uses a random partitioning strategy to create isolation trees. At each step of constructing a tree, a random feature is selected, and a random split value within the range of the selected feature's values is chosen to create a node. This random partitioning leads to shorter paths for anomalies, making them easier to isolate from normal data points.

- **Path length**: The key idea behind Isolation Forest is that anomalies are isolated into smaller partitions with fewer data points, while normal data points are distributed more uniformly across larger partitions. The average path length of a data point to reach an anomaly in a tree is used as a measure of its "isolation."

- **Anomaly score**: Based on the average path length, each data point is assigned an anomaly score. The anomaly score represents how easily the data point can be isolated or separated from the rest of the data. Shorter average path lengths correspond to higher anomaly scores, indicating that the data point is more likely to be an anomaly.

- **Contamination parameter**: The Isolation Forest algorithm has a hyperparameter called "contamination" that represents the expected proportion of anomalies in the dataset. This parameter helps in setting a threshold for identifying anomalies. The contamination parameter can be set explicitly or as "`auto`," which estimates the contamination based on the dataset's size.

A few advantages of Isolation Forest are listed here:

- Isolation Forest is computationally efficient and scalable, making it suitable for large datasets. It does not require a large number of trees to achieve good performance, reducing the computational overhead.

- The algorithm is relatively insensitive to the number of dimensions/features, which is particularly advantageous when dealing with high-dimensional datasets. Isolation Forest is an unsupervised learning algorithm, making it suitable for scenarios where labeled anomaly data is scarce or unavailable.

There are, however, some limitations of Isolation Forest:

- Isolation Forest may not perform well on datasets with multiple clusters of anomalies or when anomalies are close to the majority of normal data points.

- As with most unsupervised algorithms, Isolation Forest may produce false positives (normal data points misclassified as anomalies) and false negatives (anomalies misclassified as normal data points).

Let's implement this approach in Python with the following steps:

1. **Import libraries**: The code begins by importing the necessary libraries. The `pandas` library is imported as `pd` to handle data in tabular format, and `IsolationForest` is imported from the `sklearn.ensemble` module for performing anomaly detection using the Isolation Forest algorithm:

```
Import pandas as pd
from sklearn.ensemble import IsolationForest
```

2. **Extract numerical features**: A `numerical_features` list is defined, containing the names of the numerical columns to be used for anomaly detection. These columns are `'ApplicantIncome'`, `'CoapplicantIncome'`, and `'LoanAmount'`:

```
numerical_features = ['ApplicantIncome', 'CoapplicantIncome',
'LoanAmount']
```

3. **Create a DataFrame for anomaly detection**: A new `X_anomaly` DataFrame is created by extracting the columns specified in `numerical_features` from the original `df` DataFrame. This new DataFrame will be used for anomaly detection:

```
X_anomaly = df[numerical_features]
```

4. **Handle missing values**: To handle any missing values in the `X_anomaly` DataFrame, the `fillna()` method is used with the mean of each column. This ensures that any missing values are replaced with the mean value of their respective columns:

```
X_anomaly.fillna(X_anomaly.mean(), inplace=True)
```

5. **Initialize the Isolation Forest model**: The Isolation Forest model is initialized with `IsolationForest(contamination='auto', random_state=42)`. The `contamination` parameter is set to `'auto'`, which means it will automatically detect the percentage of outliers in the dataset. The `random_state` parameter is set to `42` to ensure reproducibility:

```
Isolation_forest = IsolationForest(contamination='auto', random_
state=42)
```

6. **Fit the model and predict anomalies**: The Isolation Forest model is fitted to the `X_anomaly` data using the `fit_predict()` method. This method simultaneously fits the model to the data and predicts whether each data point is an outlier or not. The predictions are stored in the `anomaly_predictions` array.

7. **Add anomaly predictions to the original dataset**: The `anomaly_predictions` array contains predicted labels for each data point: `-1` for anomalies (outliers) and `1` for inliers (non-outliers). These predictions are added as a new `'IsAnomaly'` column to the original `df` DataFrame.

8. **Display rows with anomalies**: Finally, the code filters rows in the `df` DataFrame where `IsAnomaly` is equal to -1, indicating the presence of outliers. The resulting DataFrame contains all rows with anomalies, which can then be further analyzed or processed as needed:

```
anomaly_predictions = Isolation_forest.fit_predict(X_anomaly)
df['IsAnomaly'] = anomaly_predictions
anomalies = df[df['IsAnomaly'] == -1]
anomalies.head()
```

We can now view the DataFrame with rows that the model has predicted as anomalies:

	Loan_ID	Gender	Married	Dependents	Education	Self_Employed	ApplicantIncome	CoapplicantIncome	LoanAmount	Loan_Amount_Term	Credit_Hi
9	LP001020	Male	Yes	1	Graduate	No	12841	10968.0	349.0	360.0	
12	LP001028	Male	Yes	2	Graduate	No	3073	8106.0	200.0	360.0	
14	LP001030	Male	Yes	2	Graduate	No	1299	1086.0	17.0	120.0	
21	LP001046	Male	Yes	1	Graduate	No	5955	5625.0	315.0	360.0	
34	LP001100	Male	No	3+	Graduate	No	12500	3000.0	320.0	360.0	
106	LP001369	Male	Yes	2	Graduate	No	11417	1126.0	225.0	360.0	
122	LP001431	Female	No	0	Graduate	No	2137	8980.0	137.0	360.0	
126	LP001448	NaN	Yes	3+	Graduate	No	23803	0.0	370.0	360.0	
128	LP001451	Male	Yes	1	Graduate	Yes	10513	3850.0	160.0	180.0	
130	LP001469	Male	No	0	Graduate	Yes	20166	0.0	650.0	480.0	
135	LP001488	Male	Yes	3+	Graduate	No	4000	7750.0	290.0	360.0	
155	LP001536	Male	Yes	3+	Graduate	No	39999	0.0	600.0	180.0	
159	LP001552	Male	Yes	0	Graduate	No	4583	5625.0	255.0	360.0	
171	LP001585	NaN	Yes	3+	Graduate	No	51763	0.0	700.0	300.0	
177	LP001610	Male	Yes	3+	Graduate	No	5516	11300.0	495.0	360.0	

Figure 9.6 – The resulting anomalies DataFrame

In conclusion, Isolation Forest stands out as a powerful and efficient tool for anomaly detection, particularly in scenarios where anomalies are rare and distinctly different from normal instances. Its ability to isolate anomalies through the creation of random trees makes it a valuable asset in various applications, from fraud detection to network security. However, it's essential to acknowledge the algorithm's limits. Isolation Forest might face challenges when anomalies are not well separated or when datasets are highly dimensional. In the next section, we will explore autoencoders – a semi-supervised method for anomaly detection.

Semi-supervised methods using autoencoders

Anomaly detection using autoencoders is an unsupervised learning approach that leverages neural networks (NNs) to detect anomalies in data. Autoencoders are a type of NN architecture designed to reconstruct the input data from a compressed representation. In anomaly detection, we exploit the fact that autoencoders struggle to reconstruct anomalous instances, making them useful for identifying unusual patterns or outliers.

Autoencoders consist of two main components: an encoder and a decoder. The encoder compresses the input data into a lower-dimensional representation called the "latent space," while the decoder tries to reconstruct the original input from this representation. The encoder and decoder are typically symmetric, and the network is trained to minimize reconstruction errors.

In anomaly detection, we train the autoencoder on normal data without anomalies. Since the autoencoder learns to reconstruct normal data, it will be less capable of reconstructing anomalies, leading to higher reconstruction errors for anomalous instances. This property allows us to use the reconstruction error as an anomaly score.

During training, we compare the original input (for example, numerical features) to the reconstructed output. The difference between the two is the reconstruction error. A low reconstruction error indicates that the input is close to the normal data distribution, while a high reconstruction error suggests that the input is likely an anomaly.

After training the autoencoder, we need to set a threshold to distinguish between normal and anomalous instances based on the reconstruction error. There are several methods to set the threshold, such as percentile-based or using validation data. The threshold will depend on the desired trade-off between false positives and false negatives, which can be adjusted based on the application's requirements.

Autoencoders are flexible and can capture complex patterns in the data, making them suitable for high-dimensional data with non-linear relationships. They can handle both global and local anomalies, meaning they can detect anomalies that differ from the majority of data points and anomalies within specific regions of the data. Autoencoders are capable of unsupervised learning, which is advantageous when labeled anomaly data is limited or unavailable. As with other unsupervised methods, autoencoders may produce false positives (normal data misclassified as anomalies) and false negatives (anomalies misclassified as normal data). They may struggle to detect anomalies that are very similar to the normal data, as the reconstruction error might not be significantly different.

Let's see an example implementation of this approach using the TensorFlow library and the Loan Prediction dataset in Python:

- **Load data**: The code begins by importing necessary libraries and loading the dataset from the `train_loan_prediction.csv` file using `pd.read_csv()`:

```
import pandas as pd
import numpy as np
from sklearn.model_selection import train_test_split
from sklearn.preprocessing import StandardScaler
from tensorflow.keras.layers import Input, Dense
from tensorflow.keras.models import Model


df = pd.read_csv('train_loan_prediction.csv')
```

- **Extract numerical features**: The code defines a `numerical_features` list containing the names of the numerical columns to be used for anomaly detection. These columns are `'ApplicantIncome'`, `'CoapplicantIncome'`, and `'LoanAmount'`:

```
numerical_features = ['ApplicantIncome', 'CoapplicantIncome',
'LoanAmount']
```

- **Create a DataFrame for anomaly detection**: A new `X_anomaly` DataFrame is created by extracting the columns specified in `numerical_features` from the original `df` DataFrame. This new DataFrame will be used for anomaly detection.

- **Handle missing values**: Any missing values in the `X_anomaly` DataFrame are replaced with the mean value of their respective columns using the `fillna()` method:

```
X_anomaly = df[numerical_features]
X_anomaly.fillna(X_anomaly.mean(), inplace=True)
```

- **Standardize numerical features**: Numerical features in `X_anomaly` are standardized using `StandardScaler()`. Standardization scales the features to have zero mean and unit variance, which is important for training machine learning models:

```
original_indices = X_anomaly.index
scaler = StandardScaler()
X_anomaly_scaled = scaler.fit_transform(X_anomaly)
```

- **Split data into training and testing sets**: The standardized data is split into training (`X_train`) and testing (`X_test`) sets using the `train_test_split()` function. The original indices of the data are also stored in `original_indices`:

```
X_train, X_test, _, _ = train_test_split(X_anomaly_scaled,
original_indices, test_size=0.2, random_state=42)
X_test_df = pd.DataFrame(X_test, columns=['ApplicantIncome',
'CoapplicantIncome', 'LoanAmount'])
```

- **Build and train an autoencoder model**: An autoencoder neural network model is constructed using TensorFlow's Keras API. The autoencoder is an unsupervised learning model designed to reconstruct the input data. It consists of an encoder and decoder, both composed of `Dense` layers. The model is trained using the `fit()` method, with the **mean squared error** (**MSE**) as the loss function:

```
input_dim = X_anomaly.shape[1]
encoding_dim = 2
input_layer = Input(shape=(input_dim,))
encoder_layer = Dense(encoding_dim, activation='relu')(input_
layer)
decoder_layer = Dense(input_dim, activation='sigmoid')(encoder_
layer)
autoencoder = Model(inputs=input_layer, outputs=decoder_layer)
autoencoder.compile(optimizer='adam', loss='mean_squared_error')
autoencoder.fit(X_anomaly_scaled, X_anomaly_scaled, epochs=50,
batch_size=16)
```

- **Reconstruct data and calculate reconstruction error**: The trained autoencoder is used to reconstruct data points in `X_test`, and reconstruction errors are calculated as the mean squared difference between the original and reconstructed data:

```
X_test_reconstructed = autoencoder.predict(X_test)
reconstruction_error_test = np.mean(np.square(X_test - X_test_
reconstructed), axis=1)
```

- **Define threshold for anomaly detection**: A threshold for anomaly detection is defined by calculating the 95th percentile of reconstruction errors in `X_test`:

```
threshold = np.percentile(reconstruction_error_test, 95)
```

- **Predict anomalies**: Anomalies are predicted by comparing reconstruction errors against the threshold. If the reconstruction error for a data point is greater than the threshold, it is classified as an anomaly and assigned a value of `1` in `anomaly_predictions`; otherwise, it is assigned `0`:

```
anomaly_predictions = (reconstruction_error_test > threshold).
astype(int)
```

- **Create a new DataFrame with anomaly predictions**: A new `anomaly_df` DataFrame is created with the anomaly predictions and the corresponding index from `X_test_df`:

```
anomaly_df = pd.DataFrame({'IsAnomaly': anomaly_predictions},
index=X_test_df.index)
```

- **Merge anomaly predictions with the original DataFrame**: The anomaly predictions are merged with the original df DataFrame using the merge() method, adding the 'IsAnomaly' column to df.

- **Display rows with anomalies**: The code checks if the 'IsAnomaly' column is present in df. If present, it displays rows where 'IsAnomaly' is equal to 1, indicating the presence of anomalies. If not present, it prints "No anomalies detected.":

```
df = df.merge(anomaly_df, how='left', left_index=True, right_
index=True)

if 'IsAnomaly' in df.columns:
    # Display the rows with anomalies
    anomalies = df[df['IsAnomaly'] == 1]
    anomalies
else:
    print("No anomalies detected.")
```

The resulting DataFrame is as follows:

	Loan_ID	Gender	Married	Dependents	Education	Self_Employed	ApplicantIncome	CoapplicantIncome	LoanAmount	Loan_Amount_Term
24	LP001052	Male	Yes	1	Graduate	NaN	3717	2925.0	151.0	360.0
60	LP001205	Male	Yes	0	Graduate	No	2500	3796.0	120.0	360.0
67	LP001233	Male	Yes	1	Graduate	No	10750	0.0	312.0	360.0
89	LP001310	Male	Yes	0	Graduate	No	5695	4167.0	175.0	360.0
90	LP001316	Male	Yes	0	Graduate	No	2958	2900.0	131.0	360.0
104	LP001357	Male	NaN	NaN	Graduate	No	3816	754.0	160.0	360.0
122	LP001431	Female	No	0	Graduate	No	2137	8980.0	137.0	360.0

Figure 9.7 – The resulting IsAnomaly DataFrame

In summary, autoencoders prove to be a versatile and powerful tool for anomaly detection, capturing nuanced patterns that may elude traditional methods. Their ability to discover subtle anomalies within complex data structures makes them invaluable in diverse domains, including image analysis, cybersecurity, and industrial quality control.

However, the effectiveness of autoencoders is contingent on various factors. The architecture's complexity and the selection of hyperparameters can influence performance, requiring careful tuning for optimal results. In the next section, we will understand how SVMs can be used in anomaly detection.

Supervised methods using SVMs

SVMs are a powerful class of supervised learning algorithms commonly used for classification tasks. When applied to anomaly detection, SVMs prove to be effective in separating normal instances from anomalies by finding a hyperplane with a maximum margin. Here is how SVMs work under the hood:

- **Hyperplane definition**: In a two-dimensional space, a hyperplane is a flat, two-dimensional subspace. SVM aims to find a hyperplane that best separates the dataset into two classes — normal and anomalous. This hyperplane is positioned to maximize the margin, which is the distance between the hyperplane and the nearest data points of each class.

- **Decision boundary**: The hyperplane serves as a decision boundary that separates instances of one class from another. In a binary classification scenario, instances on one side of the hyperplane are classified as belonging to one class, and those on the other side are classified as belonging to the other class.

- **Kernel trick**: SVM can handle complex relationships in the data through the use of a kernel function. In many real-world scenarios, the relationship between features may not be linear. SVM addresses this by using a kernel function. This function transforms the input data into a higher-dimensional space, making it easier to find a hyperplane that effectively separates the classes. Commonly used kernel functions include the linear kernel (for linearly separable data), polynomial kernel, **radial basis function** (**RBF**) or Gaussian kernel, and sigmoid kernel. The choice of kernel depends on the nature of the data.

- **Optimal hyperplane**: SVM aims to find the hyperplane that maximizes the margin, which is the distance between the hyperplane and the nearest data points of each class. The larger the margin, the more robust and generalizable the model is likely to be. Support vectors are data points that lie closest to the decision boundary. They play a crucial role in defining the optimal hyperplane and the margin. SVM focuses on these support vectors during training.

Let's implement a Python example of SVM in anomaly detection using our Loan Prediction dataset:

1. **Load data**: Let's begin by importing the necessary libraries and loading the dataset from the `train_loan_prediction.csv` file using `pd.read_csv()`:

```
import pandas as pd
from sklearn.model_selection import train_test_split
from sklearn.preprocessing import StandardScaler
from sklearn.svm import OneClassSVM
from sklearn.metrics import classification_report, accuracy_
score
df = pd.read_csv('train_loan_prediction.csv')
```

2. **Data preprocessing**: We will do some basic data preprocessing tasks, including handling missing values. For this anomaly detection example, we simplify the analysis by excluding categorical variables. In a more complex analysis, you might choose to encode and include these variables if they are deemed relevant to your specific anomaly detection task:

```
df = df.drop(['Loan_ID', 'Gender', 'Married', 'Dependents',
'Education', 'Self_Employed', 'Property_Area'], axis=1)
df['Loan_Status'] = df['Loan_Status'].map({'Y': 0, 'N': 1})
df.fillna(df.mean(), inplace=True)
```

3. Create a train-test split for the SVM model:

```
X = df.drop('Loan_Status', axis=1)
y = df['Loan_Status']
X_train, X_test, y_train, y_test = train_test_split(X, y, test_
size=0.2, random_state=42)
```

4. Standardize the features using `StandardScaler` from scikit-learn to ensure that all features have the same scale:

```
scaler = StandardScaler()
X_train_scaled = scaler.fit_transform(X_train)
X_test_scaled = scaler.transform(X_test)
```

5. Train the One-Class SVM model for anomaly detection. Adjust the *nu* parameter based on the expected proportion of outliers in your dataset. The *nu* parameter represents an upper bound on the fraction of margin errors and a lower bound on the fraction of support vectors. It essentially controls the proportion of outliers or anomalies the algorithm should consider. Choosing an appropriate value for *nu* is crucial, and it depends on the characteristics of your dataset and the expected proportion of anomalies. Here are some guidelines to help you select the *nu* parameter:

    *   **Understand the nature of anomalies**: Assess the domain knowledge and characteristics of your dataset. Understand the expected proportion of anomalies. If anomalies are rare, a smaller value of *nu* might be appropriate.

    *   **Experiment with a range of values**: Start by experimenting with a range of *nu* values, such as 0.01, 0.05, 0.1, 0.2, and so on. You can adjust this range based on your understanding of the data.

    *   **Consider the dataset size**: The size of your dataset can also influence the choice of *nu*. For larger datasets, a smaller value might be suitable, while for smaller datasets, a relatively larger value may be appropriate.

    *   **Balance false positives and false negatives**: Depending on the application, you might prioritize minimizing false positives or false negatives. Adjust *nu* accordingly to achieve the desired balance.

6. We will implement an experiment with a range of values for *nu*. We will specify a list of *nu* values that we want to experiment with. These values represent the upper bound of the fraction of margin errors and the lower bound of the fraction of support vectors in the One-Class SVM model. We then create an empty list to store the mean decision function values for each *nu* value:

```
nu_values = [0.01, 0.05, 0.1, 0.2, 0.3]
mean_decision_function_values = []
```

7. For each *nu* value in the list, train a One-Class SVM model with that *nu* value. Retrieve the decision function values for the test set and calculate the mean decision function value. Append the mean decision function value to the list:

```
for nu in nu_values:
    svm_model = OneClassSVM(nu=nu, kernel='rbf', gamma=0.1)
    svm_model.fit(X_train_scaled)
    decision_function_values=
    svm_model.decision_function(X_test_scaled)
    mean_decision_function = np.mean(decision_function_values)
    mean_decision_function_values.append(mean_decision_function)
```

8. Identify the index of the *nu* value that corresponds to the highest mean decision function value. Then, retrieve the best *nu* value:

```
best_nu_index = np.argmax(mean_decision_function_values)
best_nu = nu_values[best_nu_index]
```

9. Create a final One-Class SVM model using the best *nu* value and train it on the scaled training data:

```
final_model = OneClassSVM(nu=best_nu, kernel='rbf', gamma=0.1)
final_model.fit(X_train_scaled)
```

10. We now use this model to predict anomalies on the `X_test_scaled` test dataset. This line creates a binary representation of the predictions (`y_pred`) by mapping -1 to 1 (indicating anomalies) and any other value (typically 1) to 0 (indicating normal instances). This is done because the One-Class SVM model often assigns -1 to anomalies and 1 to normal instances. We will store this in a new DataFrame as `df_with_anomalies`:

```
y_pred = final_model.predict(X_test_scaled)
y_pred_binary = [1 if pred == -1 else 0 for pred in y_pred]
test_set_df = pd.DataFrame(data=X_test_scaled, columns=X.
columns, index=X_test.index)
test_set_df['Anomaly_Label'] = y_pred_binary
df_with_anomalies = pd.concat([df, test_set_df['Anomaly_
Label']], axis=1, join='outer')
df_with_anomalies['Anomaly_Label'].fillna(0, inplace=True)
```

11. Print the confusion matrix and accuracy score:

```
print("Classification Report:\n", classification_report(y_test,
y_pred_binary))print("Accuracy Score:", accuracy_score(y_test,
y_pred_binary))
```

This will print the following report:

```
Classification Report:
              precision    recall  f1-score   support

           0       0.65      0.97      0.78        80
           1       0.33      0.02      0.04        43

    accuracy                           0.64       123
   macro avg       0.49      0.50      0.41       123
weighted avg       0.54      0.64      0.52       123

Accuracy Score: 0.6422764227642277
```

Figure 9.8 – Output classification report

12. Print DataFrame rows predicted as anomalies:

```
df_with_anomalies[df_with_anomalies['Anomaly_Label'] == 1]
```

This will output the following DataFrame:

	ApplicantIncome	CoapplicantIncome	LoanAmount	Loan_Amount_Term	Credit_History	Loan_Status	Anomaly_Label
155	39999	0.0	600.0	180.0	0.0	0	1.0
333	63337	0.0	490.0	180.0	1.0	0	1.0
568	2378	0.0	9.0	360.0	1.0	1	1.0

Figure 9.9 – The df_with_anomalies DataFrame

In this section, we've walked through the process of implementing anomaly detection using SVM in Python. Anomaly detection using SVM can be adapted for various datasets with clear anomalies, making it a valuable tool for outlier identification. In the next section, we will explore data augmentation and resampling techniques for identifying edge cases and rare events.

Data augmentation and resampling techniques

Class imbalance is a common issue in datasets with rare events. Class imbalance can adversely affect the model's performance, as the model tends to be biased toward the majority class. To address this, we will explore two resampling techniques:

- **Oversampling**: Increasing the number of instances in the minority class by generating synthetic samples
- **Undersampling**: Reducing the number of instances in the majority class to balance class distribution

Let's discuss these resampling techniques in more detail.

Oversampling using SMOTE

Synthetic Minority Over-sampling TEchnique (**SMOTE**) is a widely used resampling method for addressing class imbalance in machine learning datasets, especially when dealing with rare events or minority classes. SMOTE helps to generate synthetic samples for the minority class by interpolating between existing minority class samples. This technique aims to balance class distribution by creating additional synthetic instances, thereby mitigating the effects of class imbalance. In a dataset with class imbalance, the minority class contains significantly fewer instances than the majority class. This can lead to biased model training, where the model tends to favor the majority class and performs poorly on the minority class.

Here are the key steps of a SMOTE algorithm:

1. **Identifying minority class instances**: The first step of SMOTE is to identify instances belonging to the minority class.

2. **Selecting nearest neighbors**: For each minority class instance, SMOTE selects its k nearest neighbors (commonly chosen through the **k-nearest neighbors** (**KNN**) algorithm). These neighbors are used to create synthetic samples.

3. **Creating synthetic samples**: For each minority class instance, SMOTE generates synthetic samples along the line connecting the instance to its k nearest neighbors in the feature space. Synthetic samples are created by adding a random fraction (usually between 0 and 1) of the feature differences between the instance and its neighbors. This process effectively introduces variability to synthetic samples.

4. **Combining with the original data**: The synthetic samples are combined with the original minority class instances, resulting in a resampled dataset with a more balanced class distribution.

SMOTE helps to address class imbalance without discarding any data, as it generates synthetic samples rather than removing instances from the majority class. It increases the information available to the model, potentially improving the model's ability to generalize to the minority class. SMOTE is straightforward to implement and is available in popular libraries such as imbalanced-learn in Python.

While SMOTE is effective in many cases, it might not always perform optimally for highly imbalanced datasets or datasets with complex decision boundaries. Generating too many synthetic samples can lead to overfitting on the training data, so it is crucial to choose an appropriate value for the number of nearest neighbors (k). SMOTE may introduce some noise and may not be as effective if the minority class is too sparse or scattered in the feature space. SMOTE can be combined with other techniques, such as undersampling the majority class or using different resampling ratios, to achieve better performance. It is essential to evaluate the model's performance on appropriate metrics (for example, precision, recall, or F1-score) to assess the impact of SMOTE and other techniques on the model's ability to detect rare events.

Let's implement this approach in Python:

- **Load data**: The code starts by importing necessary libraries and loading the Loan Prediction dataset from the `train_loan_prediction.csv` file using `pd.read_csv()`:

```
import pandas as pd
from imblearn.over_sampling import SMOTE
from sklearn.model_selection import train_test_split
from sklearn.metrics import classification_report
from sklearn.ensemble import RandomForestClassifier
df = pd.read_csv('train_loan_prediction.csv')
```

- **Map target variable**: The `'Loan_Status'` column in the dataset contains `'Y'` and `'N'` categorical values, which represent loan approval (`'Y'`) and rejection (`'N'`). To convert this categorical target variable into a numerical format, `'Y'` is mapped to 1 and `'N'` is mapped to 0 using the `map()` function:

```
df['Loan_Status'] = df['Loan_Status'].map({'Y': 1, 'N': 0})
```

- **Handle missing values**: The code applies mean imputation to all columns with missing values in the dataset using the `fillna()` method. This ensures that any missing values in the dataset are replaced with the mean value of their respective columns:

```
df.fillna(df.mean(), inplace=True)
```

- **Exclude non-numerical columns**: The code selects only numerical columns from the dataset to build an X feature set. The `select_dtypes()` method is used to include only columns with `float` and `int` data types while excluding non-numerical columns such as `'Loan_Status'` and `'Loan_ID'`. The y target variable is set to `'Loan_Status'`:

```
numerical_columns = df.select_dtypes(include=[float, int]).
columns
X = df[numerical_columns].drop('Loan_Status', axis=1)
y = df['Loan_Status']
```

- **Split data into training and testing sets**: The data is split into training (X_train, y_train) and testing (X_test, y_test) sets using the `train_test_split()` function from scikit-learn. The training set consists of 80% of the data, while the testing set contains 20% of the data. The `random_state` parameter is set to 42 to ensure reproducibility:

```
X_train, X_test, y_train, y_test = train_test_split(X, y, test_
size=0.2, random_state=42)
```

- **Instantiate SMOTE**: Here, SMOTE is instantiated with SMOTE(random_state=42). SMOTE is then applied to the training data using the fit_resample() method. This method oversamples the minority class (loan rejection) by generating synthetic samples, creating a balanced dataset. The resulting resampled data is stored in X_train_resampled and y_train_resampled:

```
smote = SMOTE(random_state=42)
X_train_resampled, y_train_resampled = smote.fit_resample(X_
train, y_train)
```

- **Train Random Forest classifier**: A Random Forest classifier is instantiated with RandomForestClassifier(random_state=42). The classifier is trained on the resampled data (X_train_resampled, y_train_resampled) using the fit() method. The trained classifier is used to make predictions on the test data (X_test) using the predict() method. Predictions are stored in y_pred:

```
clf = RandomForestClassifier(random_state=42)
clf.fit(X_train_resampled, y_train_resampled)
y_pred = clf.predict(X_test)
```

- A classification report is generated using the classification_report() function from scikit-learn. The classification report provides precision, recall, F1-score, and support for each class (loan approval and rejection) based on predictions (y_pred) and true labels (y_test) from the test set. The classification report is initially returned in a dictionary format. The code converts this dictionary to a clf_report DataFrame using pd.DataFrame(), making it easier to work with the data. The clf_report DataFrame is transposed using the .T attribute to have classes (0 and 1) as rows and evaluation metrics (precision, recall, F1-score, and support) as columns. This transposition provides a more convenient and readable format for further analysis or presentation:

```
clf_report = pd.DataFrame(classification_report(y_test, y_pred,
output_dict=True))
clf_report = clf_report.T
clf_report
```

Let's print the classification report to understand how good the model is:

	precision	recall	f1-score	support
0	0.733333	0.511628	0.602740	43.000000
1	0.774194	0.900000	0.832370	80.000000
accuracy	0.764228	0.764228	0.764228	0.764228
macro avg	0.753763	0.705814	0.717555	123.000000
weighted avg	0.759909	0.764228	0.752093	123.000000

Figure 9.10 – Classification report generated by the preceding code

The report indicates that the model performs moderately well in identifying instances of class 0, with a precision of 73.33%. However, recall is relatively lower at 51.16%, indicating that the model might miss some actual instances of class 0. The model excels in identifying instances of class 1, with a high precision of 77.42% and a very high recall of 90.00%. The weighted average metrics consider the class imbalance, providing a balanced evaluation across both classes. The overall accuracy of the model is 76.42%, indicating the percentage of correctly predicted instances.

In the next section, let's explore undersampling – another method to address class imbalance problems in machine learning.

Undersampling using RandomUnderSampler

Handling class imbalance with RandomUnderSampler is an effective approach to address the challenge of imbalanced datasets, where one class significantly outweighs the other class(es). In such cases, traditional machine learning algorithms may struggle to learn from the data and tend to be biased toward the majority class, leading to poor performance on the minority class or rare events.

RandomUnderSampler is a resampling technique that aims to balance class distribution by randomly removing instances from the majority class until class proportions become more balanced. By reducing the number of instances in the majority class, RandomUnderSampler ensures that the minority class is represented more proportionally, making it easier for the model to detect and learn patterns related to rare events.

Here are some key points about handling class imbalance with RandomUnderSampler:

- **Resampling for class balance**: RandomUnderSampler is a type of data-level resampling method. Data-level resampling techniques involve manipulating the training data to balance class distribution. In RandomUnderSampler, instances from the majority class are randomly selected and removed, resulting in a smaller dataset with a balanced class distribution.

- **Preserving minority class information**: Unlike some other undersampling techniques that merge instances or create synthetic samples, `RandomUnderSampler` directly removes instances from the majority class without altering minority class instances. This approach helps preserve information from the minority class, making it easier for the model to focus on learning patterns associated with rare events.

- **Potential information loss**: One potential drawback of `RandomUnderSampler` is the loss of information from the majority class. By removing instances randomly, some informative instances may be discarded, potentially leading to a reduction in the model's ability to generalize on the majority class.

- **Computationally efficient**: `RandomUnderSampler` is computationally efficient since it simply involves randomly removing instances from the majority class. This makes it faster compared to some other resampling methods.

- **Choosing the right resampling technique**: While `RandomUnderSampler` can be effective in certain scenarios, it might not always be the best choice, especially if the majority class contains important patterns and information. Careful consideration of the problem and dataset characteristics is crucial when selecting the appropriate resampling technique.

- **Combining with other techniques**: In practice, `RandomUnderSampler` can be used in combination with other techniques. For example, one can apply `RandomUnderSampler` first and then use `RandomOverSampler` (oversampling) to further balance class distribution. This approach helps in achieving a more balanced representation of both classes.

- **Evaluation and model selection**: When handling class imbalance, it is essential to evaluate the model's performance on relevant metrics such as precision, recall, F1-score, and **Area Under the ROC Curve (AUC-ROC)**. These metrics provide a comprehensive assessment of the model's ability to handle rare events and edge cases.

Let's implement this approach using Python:

Load data: The code begins by importing necessary libraries and loading the Loan Prediction dataset from the `train_loan_prediction.csv` file using `pd.read_csv()`:

```
import pandas as pd
from imblearn.under_sampling import RandomUnderSampler
from sklearn.model_selection import train_test_split
from sklearn.metrics import classification_report
from sklearn.ensemble import RandomForestClassifier
df = pd.read_csv('train_loan_prediction.csv')
```

The `'Loan_Status'` column in the dataset contains `'Y'` and `'N'` categorical values, which represent loan approval (`'Y'`) and rejection (`'N'`). To convert this categorical target variable into numerical format, `'Y'` is mapped to 1 and `'N'` is mapped to 0 using the `map()` function:

```
df['Loan_Status'] = df['Loan_Status'].map({'Y': 1, 'N': 0})
```

The following code applies mean imputation to all columns with missing values in the dataset using the `fillna()` method. This ensures that any missing values in the dataset are replaced with the mean value of their respective columns:

```
df.fillna(df.mean(), inplace=True)
```

The code selects only numerical columns from the dataset to build the X feature set. The `select_dtypes()` method is used to include only columns with data types `float` and `int` data types while excluding non-numerical columns such as `'Loan_Status'` and `'Loan_ID'`. The y target variable is set to `'Loan_Status'`:

```
numerical_columns = df.select_dtypes(include=[float, int]).columns
X = df[numerical_columns].drop('Loan_Status', axis=1)
y = df['Loan_Status']
```

The data is split into training (X_train, y_train) and testing (X_test, y_test) sets using the `train_test_split()` function from scikit-learn. The training set consists of 80% of the data, while the testing set contains 20% of the data. The `random_state` parameter is set to 42 to ensure reproducibility:

```
X_train, X_test, y_train, y_test = train_test_split(X, y, test_
size=0.2, random_state=42)
```

We then instantiate `RandomUnderSampler` and apply it to the training data using the `fit_resample()` method. This method undersamples the majority class (loan approval) to create a balanced dataset. The resulting resampled data is stored in X_train_resampled and y_train_resampled:

```
rus = RandomUnderSampler(random_state=42)
X_train_resampled, y_train_resampled = rus.fit_resample(X_train, y_
train)
```

A Random Forest classifier is then trained on the resampled data (X_train_resampled, y_train_resampled) using the `fit()` method. The trained classifier is used to make predictions on the test data (X_test) using the `predict()` method. Predictions are stored in y_pred:

```
clf = RandomForestClassifier(random_state=42)
clf.fit(X_train_resampled, y_train_resampled)
y_pred = clf.predict(X_test)
clf_report = pd.DataFrame(classification_report(y_test, y_pred,
output_dict=True))
clf_report = clf_report.T
clf_report
```

Let's inspect the classification report to assess the model performance:

	precision	recall	f1-score	support
0	0.681818	0.697674	0.689655	43.000000
1	0.835443	0.825000	0.830189	80.000000
accuracy	0.780488	0.780488	0.780488	0.780488
macro avg	0.758631	0.761337	0.759922	123.000000
weighted avg	0.781737	0.780488	0.781059	123.000000

Figure 9.11 – Classification report of the model's performance

The model shows decent performance for both classes, with higher precision, recall, and F1-score for class 1 compared to class 0. The weighted average considers the imbalance in class distribution, providing a more representative measure of overall performance. The accuracy score of 0.7805% suggests that the model correctly predicted the class for approximately 78% of instances in the test set.

In the next section, let's understand cost-sensitive learning and explore its crucial role in scenarios where rare events bear significant consequences.

Cost-sensitive learning

Cost-sensitive learning is a machine learning approach that takes into account costs associated with misclassifications of different classes during the model training process. In traditional machine learning, the focus is on maximizing overall accuracy, but in many real-world scenarios, misclassifying certain classes can have more severe consequences than misclassifying others.

For example, in a medical diagnosis application, misdiagnosing a severe disease as not present (false negative) could have more significant consequences than misdiagnosing a mild condition as present (false positive). In fraud detection, incorrectly flagging a legitimate transaction as fraudulent (false positive) might inconvenience the customer, while failing to detect actual fraudulent transactions (false negative) could lead to significant financial losses.

Cost-sensitive learning addresses these imbalances in costs by assigning different misclassification costs to different classes. By incorporating these costs into the training process, the model is encouraged to prioritize minimizing the overall misclassification cost rather than simply optimizing accuracy.

There are several approaches to implementing cost-sensitive learning:

- **Modifying loss functions**: The loss function used during model training can be modified to incorporate class-specific misclassification costs. The goal is to minimize the expected cost, which is a combination of misclassification costs and the model's predictions.

- **Class weights**: Another approach is to assign higher weights to the minority class or the class with higher misclassification costs. This technique can be applied to various classifiers, such as decision trees, random forests, and SVMs, to emphasize learning from the minority class.

- **Sampling techniques**: In addition to assigning weights, resampling techniques such as oversampling the minority class or undersampling the majority class can also be used to balance class distribution and improve the model's ability to learn from rare events.

- **Threshold adjustment**: By adjusting the classification threshold, we can control the trade-off between precision and recall, allowing us to make predictions that are more sensitive to the minority class.

- **Ensemble methods**: Ensemble methods such as cost-sensitive boosting combine multiple models to focus on hard-to-classify instances and assign higher weights to misclassified samples.

Cost-sensitive learning is especially important in scenarios where the class imbalance is severe and the consequences of misclassification are critical. By taking into account costs associated with different classes, the model can make more informed decisions and improve overall performance in detecting rare events and handling edge cases.

It is important to note that cost-sensitive learning requires careful consideration of the cost matrix, as incorrectly specified costs can lead to unintended results. Proper validation and evaluation of the model on relevant metrics, considering real-world costs, are crucial to ensure the effectiveness and reliability of cost-sensitive learning algorithms.

Let's now demonstrate cost-sensitive learning using the Loan Prediction dataset in Python:

1. Load the required libraries and datasets using `pandas`:

```
import pandas as pd
from sklearn.model_selection import train_test_split
from sklearn.ensemble import RandomForestClassifier
from sklearn.metrics import classification_report
df = pd.read_csv('train_loan_prediction.csv')
```

2. We now need to perform data preprocessing to handle missing values and convert target variables to numeric data types:

```
df['Loan_Status'] = df['Loan_Status'].map({'Y': 1, 'N': 0})
df.fillna(df.mean(), inplace=True)
```

3. For this example, we will use only numeric columns. We will then split the dataset into `train` and `test`:

```
numerical_columns = df.select_dtypes(include=[float, int]).
columns
X = df[numerical_columns].drop('Loan_Status', axis=1)
y = df['Loan_Status']
```

```
X_train, X_test, y_train, y_test = train_test_split(X, y, test_
size=0.2, random_state=42)
```

4. We first calculate the class weights based on the inverse of class frequencies in the training data. The higher the frequency of a class, the lower its weight, and vice versa. This way, the model assigns higher importance to the minority class (rare events) and is more sensitive to its correct prediction:

```
class_weights = dict(1 / y_train.value_counts(normalize=True))
```

5. Next, we train the Random Forest classifier with the `class_weight` parameter set to the calculated class weights. This modification allows the classifier to consider the class weights during the training process, effectively implementing cost-sensitive learning:

```
clf = RandomForestClassifier(random_state=42, class_
weight=class_weights)
clf.fit(X_train, y_train)
```

6. After training the model, we make predictions on the test data and evaluate the classifier's performance using the classification report, which provides precision, recall, F1-score, and support for each class:

```
y_pred = clf.predict(X_test)
clf_report = pd.DataFrame(classification_report(y_test, y_pred,
output_dict=True))
clf_report = clf_report.T
clf_report
```

Let's view the classification report and assess the Random Forest classifier's performance:

	precision	recall	f1-score	support
0	0.740741	0.465116	0.571429	43.000000
1	0.760417	0.912500	0.829545	80.000000
accuracy	0.756098	0.756098	0.756098	0.756098
macro avg	0.750579	0.688808	0.700487	123.000000
weighted avg	0.753538	0.756098	0.739309	123.000000

Figure 9.12 – Classification report of the Random Forest classifier's performance

In cost-sensitive learning, you would typically define a cost matrix that quantifies misclassification costs for each class and use it to guide the model's training. The results from the classification report can help you identify areas where adjustments may be needed to align the model with specific cost considerations in your application. A cost matrix is especially useful in situations where the costs of false positives and false negatives are not equal. If the cost of false positives is higher, consider raising the decision threshold. If the cost of false negatives is higher, consider lowering the threshold.

In the next section, let's understand which evaluation metrics are used for detecting edge cases and rare events.

Choosing evaluation metrics

When dealing with edge cases and rare events in machine learning, selecting the right evaluation metrics is crucial to accurately assess the performance of the model. Traditional evaluation metrics, such as accuracy, may not be sufficient in imbalanced datasets where the class of interest (the rare event) is vastly outnumbered by the majority class. In imbalanced datasets, where the rare event is a minority class, traditional evaluation metrics such as accuracy can be misleading. For instance, if a dataset has 99% of the majority class and only 1% of the rare event, a model that predicts all instances as the majority class will still achieve an accuracy of 99%, which is deceptively high. However, such a model would be ineffective in detecting the rare event. To address this issue, we need evaluation metrics that focus on the model's performance in correctly identifying the rare event, even at the expense of a decrease in accuracy.

Here are some evaluation metrics that are more suitable for detecting edge cases and rare events:

- **Precision**: Precision measures the accuracy of positive predictions made by the model. It is the ratio of true positive (correctly predicted rare event) to the sum of true positive and false positive (incorrectly predicted rare event as the majority class). High precision indicates that the model is cautious in making positive predictions and has a low false positive rate.

- **Recall (sensitivity)**: Recall measures the proportion of true positives predicted by the model out of all actual positive instances. It is the ratio of true positive to the sum of true positive and false negative (incorrectly predicted majority class as the rare event). High recall indicates that the model is capable of capturing a significant portion of rare event instances.

- **F1-score**: The F1-score is the harmonic mean of precision and recall. It provides a balance between the two metrics and is especially useful when there is an imbalance between precision and recall. F1-score penalizes models that prioritize either precision or recall at the expense of the other.

- **Area Under the Receiver Operating Characteristic (ROC-AUC)**: ROC-AUC is a performance metric used to evaluate binary classification models. It measures the area under the ROC curve, which plots the true positive rate (recall) against the false positive rate as the classification threshold changes. A higher ROC-AUC indicates better model performance, especially in detecting rare events.

In the next section, let's delve into ensemble techniques and understand their crucial role in machine learning models, especially when dealing with data containing edge cases and rare events.

Ensemble techniques

Ensemble techniques are powerful methods used to improve the performance of machine learning models, particularly in scenarios with imbalanced datasets, rare events, and edge cases. These techniques combine multiple base models to create a more robust and accurate final prediction. Let's discuss some popular ensemble techniques.

Bagging

Bootstrap aggregating (**bagging**) is an ensemble technique that creates multiple bootstrap samples (random subsets with replacement) from the training data and trains a separate base model on each sample. The final prediction is obtained by averaging or voting the predictions of all base models. Bagging is particularly useful when dealing with high variance and complex models, as it reduces overfitting and enhances the model's generalization ability. Here are the key concepts associated with bagging:

- **Bootstrap sampling**: The bagging process begins by creating multiple random subsets of the training data through a process called bootstrap sampling. Bootstrap sampling involves randomly selecting data points from the original dataset with replacements. As a result, some data points may appear more than once in a subset, while others may be left out.

- **Base model training**: For each bootstrap sample, a base model (learner) is trained independently on that particular subset of the training data. The base models can be any machine learning algorithm, such as decision trees, random forests, or SVMs.

- **Aggregating predictions**: Once all base models are trained, they are used to make predictions on new, unseen data. For classification tasks, the final prediction is typically determined by majority voting, where the class that receives the most votes across the base models is chosen. In regression tasks, the final prediction is obtained by averaging the predictions from all base models.

Here are a few benefits of bagging:

- **Variance reduction**: Bagging helps reduce variance in the model by combining predictions from multiple models trained on different subsets of the data. This results in a more stable and robust model.

- **Overfitting prevention**: By training each base model on different subsets of the data, bagging prevents individual models from overfitting to noise in the training set.

- **Model generalization**: Bagging improves the model's generalization ability by reducing bias and variance, leading to better performance on unseen data.

- **Parallelism**: Since the base models are trained independently, bagging is amenable to parallel processing, making it computationally efficient.

Random Forest is a popular example of the bagging technique. In Random Forest, the base models are decision trees, and the predictions from multiple decision trees are combined to make the final prediction.

Here's an example of implementing bagging using Random Forest in Python on the Loan Prediction dataset:

```
import pandas as pd
from sklearn.model_selection import train_test_split
from sklearn.ensemble import RandomForestClassifier
from sklearn.metrics import classification_report

df = pd.read_csv('train_loan_prediction.csv')
df['Loan_Status'] = df['Loan_Status'].map({'Y': 1, 'N': 0})
df.fillna(df.mean(), inplace=True)
numerical_columns = df.select_dtypes(include=[float, int]).columns
X = df[numerical_columns].drop('Loan_Status', axis=1)
y = df['Loan_Status']
X_train, X_test, y_train, y_test = train_test_split(X, y, test_
size=0.2, random_state=42)
clf = RandomForestClassifier(random_state=42)
clf.fit(X_train, y_train)
y_pred = clf.predict(X_test)
clf_report = pd.DataFrame(classification_report(y_test, y_pred,
output_dict=True))
clf_report = clf_report.T
clf_report
```

In conclusion, bagging techniques offer a robust and effective strategy for handling edge cases. By aggregating predictions from multiple base models, bagging not only enhances the overall model stability but also fortifies its ability to accurately identify and address edge cases, contributing to a more resilient and reliable predictive framework.

In the next section, we will explore boosting, another method to handle edge cases and rare events.

Boosting

Boosting is an ensemble technique that builds base models sequentially, with each subsequent model focusing on misclassified instances of the previous model. It assigns higher weights to misclassified instances, thus giving more attention to rare events. Popular boosting algorithms include **Adaptive Boosting (AdaBoost)**, Gradient Boosting, and XGBoost. Boosting aims to create a strong learner by combining weak learners iteratively.

Here's how boosting works:

1. **Base model training**: Boosting starts by training a base model (also known as a weak learner) on the entire training dataset. Weak learners are usually simple models with limited predictive power, such as decision stumps (a decision tree with a single split).

2. **Weighted training**: After the first model is trained, data points that were misclassified by the model are assigned higher weights. This means that the subsequent model will pay more attention to those misclassified data points, attempting to correct their predictions.

3. **Iterative training**: Boosting follows an iterative approach. For each iteration (or boosting round), a new weak learner is trained on the updated training data with adjusted weights. The weak learners are then combined to create a strong learner, which improves its predictive performance compared to the individual weak learners.

4. **Weighted voting**: During the final prediction, the weak learners' predictions are combined with weighted voting, where models with higher accuracy have more influence on the final prediction. This allows the boosting algorithm to focus on difficult-to-classify instances and improve the model's sensitivity to rare events.

The benefits of boosting are as follows:

- **Increased accuracy**: Boosting improves the model's accuracy by focusing on the most challenging instances in the dataset and refining predictions over multiple iterations

- **Robustness**: Boosting reduces the model's sensitivity to noise and outliers in the data by iteratively adjusting weights and learning from previous mistakes

- **Model adaptation**: Boosting adapts well to different types of data and can handle complex relationships between features and the target variable

- **Ensemble diversity**: Boosting creates a diverse ensemble of weak learners, which results in better generalization and reduced overfitting

Let's look at an example of boosting using AdaBoost.

AdaBoost is a popular boosting algorithm that is commonly used in practice. In AdaBoost, the base models are typically decision stumps, and the model's weights are adjusted after each iteration to emphasize misclassified instances.

Here's an example of implementing boosting using AdaBoost in Python:

```
import pandas as pd
from sklearn.model_selection import train_test_split
from sklearn.ensemble import AdaBoostClassifier
from sklearn.metrics import classification_report
```

```
df = pd.read_csv('train_loan_prediction.csv')
df['Loan_Status'] = df['Loan_Status'].map({'Y': 1, 'N': 0})
df.fillna(df.mean(), inplace=True)

numerical_columns = df.select_dtypes(include=[float, int]).columns
X = df[numerical_columns].drop('Loan_Status', axis=1)
y = df['Loan_Status']
X_train, X_test, y_train, y_test = train_test_split(X, y, test_
size=0.2, random_state=42)
clf = AdaBoostClassifier(random_state=42)
clf.fit(X_train, y_train)
y_pred = clf.predict(X_test)
clf_report = pd.DataFrame(classification_report(y_test, y_pred,
output_dict=True))
clf_report = clf_report.T
clf_report
```

In summary, the application of boosting techniques emerges as a robust strategy for handling edge cases. Through its iterative approach, boosting empowers models to focus on instances that pose challenges, ultimately enhancing their ability to discern and accurately predict rare events.

We will now explore how to use the stacking method to detect and handle edge cases and rare events in machine learning.

Stacking

Stacking is an advanced ensemble learning technique that combines the predictions of multiple base models by training a meta-model on their outputs. Stacking aims to leverage the strengths of different base models to create a more accurate and robust final prediction. It is a form of "learning to learn" where the meta-model learns how to best combine the predictions of the base models. The base models act as "learners," and their predictions become the input features for the meta-model, which makes the final prediction. Stacking can often improve performance by capturing complementary patterns from different base models.

Here is the model methodology:

1. **Base model training**: The stacking process starts by training multiple diverse base models on the training dataset. These base models can be different types of machine learning algorithms or even the same algorithm with different hyperparameters.

2. **Base model predictions**: Once the base models are trained, they are used to make predictions on the same training data (in-sample predictions) or a separate validation dataset (out-of-sample predictions).

3. **Meta-model training**: The predictions from the base models are then combined to create a new dataset that serves as the input for the meta-model. Each base model's predictions become a new feature in this dataset. The meta-model is trained on this new dataset along with the true target labels.

4. **Final prediction**: During the final prediction phase, the base models make predictions on the new, unseen data. These predictions are then used as input features for the meta-model, which makes the final prediction.

Stacking has the following benefits:

*   **Improved predictive performance**: Stacking leverages the complementary strengths of different base models, potentially leading to better overall predictive performance compared to using individual models

*   **Reduction of bias and variance**: Stacking can reduce the model's bias and variance by combining multiple models, leading to improved generalization

*   **Flexibility**: Stacking allows the use of diverse base models, making it suitable for various types of data and problems

*   **Ensemble diversity**: Stacking creates a diverse ensemble by using various base models, which can help prevent overfitting

Here's an example of implementing stacking using scikit-learn in Python using the Loan Prediction dataset:

1. We first import the required libraries and load the dataset:

```
import pandas as pd
from sklearn.model_selection import train_test_split
from sklearn.ensemble import RandomForestClassifier,
GradientBoostingClassifier
from sklearn.linear_model import LogisticRegression
from sklearn.metrics import classification_report
df = pd.read_csv('train_loan_prediction.csv')
```

2. We now need to perform data preprocessing to handle missing values and convert target variables to numeric data types:

```
df['Loan_Status'] = df['Loan_Status'].map({'Y': 1, 'N': 0})
df.fillna(df.mean(), inplace=True)
```

3. For simplicity, we will use only numeric columns for this example. We then split the dataset into `train` and `test`:

```
numerical_columns = df.select_dtypes(include=[float, int]).
columns
X = df[numerical_columns].drop('Loan_Status', axis=1)
```

```
y = df['Loan_Status']
X_train, X_test, y_train, y_test = train_test_split(X, y, test_
size=0.2, random_state=42)
```

4. **Instantiate and train base models**: Two base models, RandomForestClassifier and GradientBoostingClassifier, are instantiated with RandomForestClassifier(random_state=42) and GradientBoostingClassifier(random_state=42) respectively:

```
base_model_1 = RandomForestClassifier(random_state=42)
base_model_2 = GradientBoostingClassifier(random_state=42)
```

5. These base models are trained on the training data (X_train, y_train) using the fit() method, as seen next. The trained base models are used to make predictions on the test data (X_test) using the predict() method. Predictions from both base models are stored in pred_base_model_1 and pred_base_model_2:

```
base_model_1.fit(X_train, y_train)
base_model_2.fit(X_train, y_train)
pred_base_model_1 = base_model_1.predict(X_test)
pred_base_model_2 = base_model_2.predict(X_test)
```

6. **Create a new dataset for stacking**: A new stacking_X_train dataset is created by combining the predictions from the base models (pred_base_model_1 and pred_base_model_2). This new dataset will be used as input features for the meta-model:

```
stacking_X_train = pd.DataFrame({
    'BaseModel1': pred_base_model_1,
    'BaseModel2': pred_base_model_2
})
```

7. **Instantiate and train meta-model**: A meta-model (Logistic Regression, in this case) is instantiated with LogisticRegression(). The meta-model is trained on the new dataset (stacking_X_train) and the true labels from the test set (y_test) using the fit() method. The meta-model learns to combine predictions of the base models and make the final prediction:

```
meta_model = LogisticRegression()
meta_model.fit(stacking_X_train, y_test)
```

8. **Create unseen data for demonstration and make predictions**: For demonstration purposes, a new sample of unseen data (new_unseen_data) is created by randomly selecting 20% of the test data (X_test) using the sample() method. The base models are used to make predictions on the new, unseen data (new_unseen_data) using the predict() method.

Predictions from both base models for the new data are stored in `new_pred_base_model_1` and `new_pred_base_model_2`:

```
new_unseen_data = X_test.sample(frac=0.2, random_state=42)
new_pred_base_model_1 = base_model_1.predict(new_unseen_data)
new_pred_base_model_2 = base_model_2.predict(new_unseen_data)
```

9. **Create a new dataset for stacking with unseen data**: A new `stacking_new_unseen_data` dataset is created by combining predictions from the base models (`new_pred_base_model_1` and `new_pred_base_model_2`) for the new, unseen data. This new dataset will be used as input features for the meta-model to make the final prediction:

```
stacking_new_unseen_data = pd.DataFrame({
    'BaseModel1': new_pred_base_model_1,
    'BaseModel2': new_pred_base_model_2
})
```

10. **Make final prediction using meta-model**: The meta-model (Logistic Regression) is used to make the final prediction on the new, unseen data (`stacking_new_unseen_data`) using the `predict()` method. The `final_prediction` variable holds the predicted classes (0 or 1) based on the meta-model's decision:

```
final_prediction = meta_model.predict(stacking_new_unseen_data)
final_prediction
```

In summary, this code demonstrates the concept of stacking, where base models (Random Forest and Gradient Boosting) are trained on the original data, their predictions are used as input features for a meta-model (Logistic Regression), and the final prediction is made using the meta-model on new, unseen data. Stacking allows the models to work together and can potentially improve the prediction performance compared to using the base models alone.

Summary

In this chapter, we explored the critical aspect of detecting rare events and edge cases in machine learning. Rare events, by their infrequency, hold significant implications across various domains and necessitate special attention. We delved into several techniques and methodologies that equip us to effectively identify and handle these uncommon occurrences.

Statistical methods, such as Z-scores and IQR, provide powerful tools to pinpoint outliers and anomalies in our data. These methods aid in establishing meaningful thresholds for identifying rare events, enabling us to distinguish significant data points from noise.

We also explored machine learning-based anomaly detection techniques, such as isolation forest and autoencoders. These methods leverage unsupervised learning to identify patterns and deviations that diverge from the majority of the data, making them well suited for detecting rare events in complex datasets.

Additionally, we discussed the significance of resampling methods such as SMOTE and `RandomUnderSampler` to tackle class imbalances. These techniques enable us to create balanced datasets that enhance the performance of models in identifying rare events while preserving data integrity.

Furthermore, we uncovered the potential of ensemble techniques, including stacking, bagging, and boosting, in augmenting the capabilities of our models for detecting edge cases. The combined power of multiple models through ensembles enhances generalization and model robustness.

It is crucial to select appropriate evaluation metrics, especially in the presence of rare events, to ensure fair assessment and accurate model performance evaluation. Metrics such as precision, recall, F1-score, and AUC-ROC provide comprehensive insights into model performance and guide decision-making.

Detecting rare events and edge cases has far-reaching implications across diverse domains, including medical diagnosis, fraud detection, predictive maintenance, and environmental monitoring. By employing effective techniques to identify and handle these infrequent occurrences, we enhance the reliability and efficiency of machine learning applications.

As we conclude this chapter, let us recognize the significance of this skill in real-world scenarios. Detecting rare events empowers us to make informed decisions, protect against potential risks, and leverage the full potential of machine learning to drive positive impact in a multitude of fields.

In the next chapter, we will explore some of the challenges faced by the data-centric approach to machine learning.

Part 4:
Getting Started with
Data-Centric ML

By now you may have realized that shifting to a data-centric approach to ML involves not just adapting your own ways of working, but also influencing those around you – a task that's far from simple. In this part, we explore both the technical and non-technical hurdles you might encounter during the development and deployment of models, and reveal how adopting a data-centric approach can aid in overcoming these obstacles.

This part has the following chapter:

- *Chapter 10, Kick-Starting Your Journey in Data-Centric Machine Learning*

10

Kick-Starting Your Journey in Data-Centric Machine Learning

The data-centric **machine learning** (**ML**) approach has been created in response to the limitations of the model-centric paradigm. Although the data-centric perspective opens up incredible opportunities for the application of ML across new and existing domains, it doesn't mean it's easy to implement.

The appeal of the model-centric approach is its relative simplicity. Its dominance in the field isn't necessarily because it's superior, but rather because it's more straightforward. It focuses primarily on refining models, tweaking algorithms, and enhancing computational power. However, this approach often neglects a fundamental aspect of ML – the quality and relevance of the data feeding these models.

The data-centric approach, on the other hand, prioritizes improving data quality over perfecting models. It recognizes that even the most sophisticated models can falter if they're built on a shaky foundation of poorly curated or irrelevant data.

The challenge you will have to lean into is that adopting a data-centric ML approach may require a significant amount of effort.

In this concluding chapter, we will concentrate on how you can effectively implement the knowledge you've accumulated throughout this book. We will cover the following topics:

- How a data-centric approach solves six common ML challenges
- The importance of championing data quality in your organization
- Bringing people together to help you implement a data-centric approach
- Taking accountability for AI ethics and fairness

Let's start by looking at how a data-centric approach provides the tools to remove common bottlenecks in ML development.

Solving six common ML challenges

We have written this book to provide you with the tools and techniques required to overcome six common challenges that typically become bottlenecks in the development of ML models. These six challenges are as follows:

- **The typical development approach is model-centric**: In traditional ML, the focus is primarily on the model – choosing the right algorithm, tuning hyperparameters, and optimizing performance metrics. This model-centric approach often involves countless iterations of tweaking the model to squeeze out a bit more accuracy. However, while models are undoubtedly important, this approach can sometimes lead to overlooking other equally crucial aspects, such as the quality and relevancy of the data feeding these models.

- **The potential of any ML model is capped by data quality and quantity**: No matter how sophisticated an ML model is, its performance is ultimately determined by the quality and quantity of the data it is trained on. Poor quality data can lead to inaccurate predictions, while insufficient data may result in overfitting, where the model learns the training data too well and performs poorly on unseen data.

- **We can't always just "get more data"**: While having more data can help improve model performance, obtaining additional data isn't always feasible. It may be expensive, time-consuming, or even impossible due to privacy concerns or the rarity of certain events.

- **Most data is not collected and curated for ML purposes**: Data is often collected for a variety of reasons, such as reporting, **business intelligence** (**BI**), or record-keeping, but not specifically for ML. This can result in data that is irrelevant, incomplete, or noisy for our specific ML task.

- **SMEs and annotators don't know the end goal of data collection**: In many cases, those collecting or annotating the data are not fully aware of the end goal of the ML project. This lack of understanding can lead to inconsistencies in data annotation or collection, as different people might interpret instructions differently.

- **SMEs and annotators are biased**: Everyone has biases, and SMEs and annotators are no exception. These biases can inadvertently influence how they collect or annotate data, leading to skewed datasets.

To solve these six challenges, or bottlenecks, we presented you with four principles of data-centric ML (in *Chapter 3, Principles of Data-Centric ML*) and a data-centric ML toolkit consisting of a range of technical and non-technical tools and approaches.

Figure 10.1 illustrates how these principles, tools, and approaches give you the tools to overcome these challenges and achieve more with your efforts:

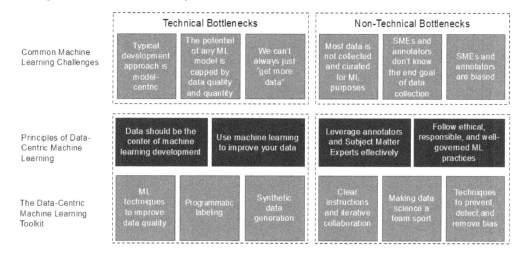

Figure 10.1 – An overview of six common challenges or bottlenecks in the development of ML models; this book provides you with four principles and a range of data-centric tools and techniques to overcome these challenges

However, knowledge is only useful when it is applied. As we draw a close to our exploration of data-centric ML, it is time for you to reflect on what you will have to do differently to apply what you have learned in this book.

Inevitably, you will have to change some of your own approaches and habits and reshape some of the interactions you have with your colleagues. As we have mentioned many times over, data science is a team sport, and it is time for you to become the captain of the team playing data-centric ML.

To successfully navigate this role, there are three key areas where you need to focus:

- Being a champion for data quality
- Bringing people together to help you
- Being accountable for AI ethics and fairness

Let's go through each of these areas in detail.

Being a champion for data quality

The cornerstone of any successful data-centric ML initiative is high-quality data. As an expert in this space, it's your responsibility to advocate for and maintain rigorous data quality standards. This involves ensuring that data collection, cleaning, and annotation processes are robust and consistent. However, it may sometimes feel like the universe is against you, continuously supplying you with poor data. That's probably true!

The second law of thermodynamics states that energy tends to spread out and systems naturally progress toward disorder or increase in "entropy" if not actively managed.

This phenomenon also plays out in pretty much every ordered system you can think of, including business. At its core, a business is a collection of people, technologies, and processes systematically organized toward a set of common goals.

But as time goes on, people will move in and out of the business, technology stacks grow in size and complexity, and processes become redundant or superseded by people's personal interests.

Let's apply this concept to a business scenario where an organization's technology stack grows over time. As the tech stack expands, adding more software, platforms, and infrastructure, it's like adding more energy to the system. Without proper management and organization, this can lead to increased complexity and "disorder." Systems may not interact efficiently, there might be redundancies, or critical processes could fall through the cracks – this is the "entropy" in our scenario.

We see this scenario play out in every organization we interact with. Even though any organization would have started from zero, the typical medium- to large-sized corporation will have hundreds, if not thousands, of technology applications running at any given time, generating an increasingly complex web of data.

A 2022 survey of more than 200 IT and data professionals at North American organizations with at least 1,000 employees confirms this[1]. The survey found that these organizations manage data from an average of 400 sources, with 20% of respondents pulling data from 1,000 or more sources to feed their analytics systems.

In other words, data disorder is not a random occurrence, but the default outcome. Unless lots of effort is put into keeping things in order, chaotic data *will* ensue. As a data professional, it is your job to identify and reverse this data entropy, with help from the rest of the organization. If you don't, your job will become harder, not easier, over time.

To manage and reverse data entropy, you first need to measure it. There are several sophisticated tools available in the marketplace for this purpose, but if your organization isn't able and willing to invest in these, then start championing data quality using simple data quality measures. If your organization doesn't have effective data quality measurement protocols in place, you can be the one to inspire action.

Championing data quality means showing others how it's done. In this book, we have given you the tools you need to take a data-centric approach to ML. Now, it's your turn to inspire your colleagues by showing them the potential of this approach. Be generous in sharing the knowledge you've gained throughout this book.

Championing data quality also means creating an environment where data accuracy is valued and prioritized. Encourage your team to never make assumptions about the state of the data. By doing so, you'll be laying the groundwork for reliable, accurate, and effective ML models.

And to do this, you need to bring people together from across the business.

Bringing people together

Data science is, at its core, a collaborative discipline. It requires the input and expertise of individuals from various backgrounds and skill sets. As the expert at the center of any ML project, one of your primary roles is to facilitate this collaboration.

Here is the challenge: most data scientists are curious, technically gifted, autonomous, and highly intelligent systematic thinkers. We like to solve complex problems, and we like to solve them *on our own*. We like working in solitude. We revel in detail. We are used to people asking *us* for help, not the other way around. We are problem solvers, not problem creators.

For people with these behavioral traits, it can be really tough to bring people together and manage the inevitable interpersonal complexities that emerge from group dynamics. It can take a lot of energy and push us beyond what we're comfortable with. Out of everything we've covered in this book, learning to bring people together might be the hardest part.

At the same time, our stakeholders typically don't have a deep technical understanding of ML. They're probably not even interested in how it all works – they just want the business outcome. They want to see results, not data quality issues. This is a tough crowd to inspire.

It's not easy to foster an environment where open communication, idea sharing, and constructive feedback are the norm. It's not easy to break down silos and encourage cross-functional collaboration. But that is what it takes if you want to maximize your chances of building impactful ML solutions that make a difference.

You get non-technical stakeholders excited about data quality by linking it to a business problem, not by describing the technical intricacies. This requires you to understand the business, more than it requires business stakeholders to understand data.

Therefore, we encourage you to step outside your comfort zone and do your very best to make data science a team sport, even if it is really difficult and feels uncomfortable. It will be very rewarding when you succeed. Remember – the most innovative solutions often come from diverse teams that bring together different perspectives and approaches.

Finally, adopting a data-centric approach extends beyond merely optimizing ML performance. It also involves the important responsibility of using data ethically and fairly.

Taking accountability for AI ethics and fairness

As the use of ML and AI continues to grow, so does the importance of ethics and fairness in these technologies. As an expert in this field, it's your responsibility to ensure that the data you build models on is as fair and unbiased as possible.

ML is an act of creation. It's a process that shapes the world around us, influencing how we interact with products, services, and even each other. ML models have the power to shape behaviors, mold perceptions, and establish norms.

This power can be used for good, creating products that enhance lives and solve pressing social issues. But it can also be misused, leading to harmful outcomes. As designers of ML solutions, it's crucial that we hold ourselves accountable for our designs and their impact on the world. It's not enough to create something that is functionally accurate or technologically advanced.

This means being proactive in identifying potential sources of bias and taking steps to mitigate them. It also means being transparent about how models are built and used, and being accountable for their outcomes. By owning this responsibility, you'll be helping to build trust in ML systems and ensuring that they are used in a way that is fair and beneficial for all.

Making data everyone's business – our own experience

A few years ago, I (Jonas) took over the leadership of a highly skilled team that included data engineers, BI developers, data analysts, and data scientists.

Although this was a very competent team with a solid understanding of their business, they struggled to live up to their full potential. I quickly identified the team's number one challenge: stakeholders from across the organization were inundating the team with basic data requests that kept everyone very busy but were not particularly impactful.

In short, stakeholders were accustomed to requesting raw or minimally curated data so that they could find their own insights, and the team was conditioned to reluctantly accept this as the modus operandi.

The scattered use of data and analytics showed that the organization hadn't grasped the importance of data quality. They were also missing out on greatly boosting business performance by using advanced analytics in a more organized way.

We wanted to flip this situation around so that we could establish a more structured, streamlined approach to data usage and analytics. Our goal was to emphasize the importance of data quality and its potential to drive enhanced business performance. We aimed to change the approach from self-service raw data to providing valuable insights through well-curated, advanced analytics, thus transforming the modus operandi from a reactive to a proactive one.

We had three areas of focus:

- We wanted to open people's eyes to the importance of data quality. Our definition of success was seeing key business stakeholders turn into self-nominated data quality champions.

- We wanted to deliver advanced data products that created enduring business value. Our definition of success was seeing our team and business stakeholders coming together to define, build, and use data-driven business tools and products.

- We wanted data to be an asset for the company and its customers. Our definition of success was that data-driven solutions delivered by the team were creating enduring value for customers and employees alike.

We started by holding companywide education sessions on the importance of data quality and the potential to improve business performance with ML. People were nodding their heads in agreement and commending the team for making this complex information digestible by a layman's audience. But we didn't see the behavior change we wanted.

The influx of basic requests amplified as our stakeholders had now been inspired to use data to inform their decisions. The team got busier than ever, trying to make the best of these poorly designed requests.

We quickly decided to change our approach. Firstly, we introduced a requirement for a business case to be established before we committed to any new piece of work that would take more than half a day to complete.

Secondly, we conducted a series of workshops with senior stakeholders to identify and rank the company's biggest pain points that could be solved through better provision of advanced analytics.

From these workshops came three major projects that involved stakeholders from across the organization: the provision of a new portfolio management system for customer-facing staff, a task management system for back-office staff, and a process automation project. All three projects relied on rules-based and ML algorithms and required an increased focus on data quality.

The projects relied on business stakeholders taking charge of data quality and creating processes that would ensure data quality always met our standards. Our team collaborated closely with business stakeholders to make sure there was always a clear link between business problems and data capture.

To our great excitement, it didn't take long before meeting agendas around the company included data quality issues and initiatives. Our senior stakeholders had taken on the role of championing data quality and were holding their teams accountable.

Each of the three projects was a resounding success, producing robust data products while simultaneously enhancing cross-functional collaboration. These projects generated a shift toward data-driven decision-making and reinforced the perception of data as a strategic asset across the company.

This significant transformation was only possible because the data professionals on my team bravely stepped outside their comfort zones to make data everybody's business.

Summary

Throughout this book, you've gained invaluable skills, and you now have the knowledge to reach the next frontier of ML, following a data-centric approach. However, this knowledge is only useful if you apply it, and that takes effort.

It requires you to use the tools and techniques outlined in this book. It also requires you to step outside your comfort zone and take the lead on data quality, cross-functional collaboration, and data ethics. You can't change the world alone, so hand over this book to a colleague and get them onboard with data centricity.

As we conclude, remember that embarking on this data-centric ML journey is not a destination but a continuous process. The landscape of data and AI is ever-evolving, and so must we evolve. Let's continue to learn, innovate, and shape the future of data science and ML together. Your data-centric journey has only just begun.

References

1. `https://www.matillion.com/blog/matillion-and-idg-survey-data-growth-is-real-and-3-other-key-findings`, viewed August 30, 2023

Index

A

B

packtpub.com

Subscribe to our online digital library for full access to over 7,000 books and videos, as well as industry leading tools to help you plan your personal development and advance your career. For more information, please visit our website.

Why subscribe?

- Spend less time learning and more time coding with practical eBooks and Videos from over 4,000 industry professionals

- Improve your learning with Skill Plans built especially for you

- Get a free eBook or video every month

- Fully searchable for easy access to vital information

- Copy and paste, print, and bookmark content

Did you know that Packt offers eBook versions of every book published, with PDF and ePub files available? You can upgrade to the eBook version at packtpub.com and as a print book customer, you are entitled to a discount on the eBook copy. Get in touch with us at customercare@packtpub.com for more details.

At www.packtpub.com, you can also read a collection of free technical articles, sign up for a range of free newsletters, and receive exclusive discounts and offers on Packt books and eBooks.

Other Books You May Enjoy

If you enjoyed this book, you may be interested in these other books by Packt:

Practical Machine Learning on Databricks

Debu Sinha

ISBN: 978-1-80181-203-0

- Transition smoothly from DIY setups to Databricks
- Master AutoML for quick ML experiment setup
- Automate model retraining and deployment
- Leverage Databricks feature store for data prep
- Use MLflow for effective experiment tracking
- Gain practical insights for scalable ML solutions
- Find out how to handle model drifts in production environments

MATLAB for Machine Learning

Giuseppe Ciaburro

ISBN: 978-1-83508-769-5

- Discover different ways to transform data into valuable insights
- Explore the different types of regression techniques
- Grasp the basics of classification through Naive Bayes and decision trees
- Use clustering to group data based on similarity measures
- Perform data fitting, pattern recognition, and cluster analysis
- Implement feature selection and extraction for dimensionality reduction
- Harness MATLAB tools for deep learning exploration

Packt is searching for authors like you

If you're interested in becoming an author for Packt, please visit `authors.packtpub.com` and apply today. We have worked with thousands of developers and tech professionals, just like you, to help them share their insight with the global tech community. You can make a general application, apply for a specific hot topic that we are recruiting an author for, or submit your own idea.

Share Your Thoughts

Now you've finished *Data-Centric Machine Learning with Python*, we'd love to hear your thoughts! Scan the QR code below to go straight to the Amazon review page for this book and share your feedback or leave a review on the site that you purchased it from.

`https://packt.link/r/1-804-61812-8`

Your review is important to us and the tech community and will help us make sure we're delivering excellent quality content.

Download a free PDF copy of this book

Thanks for purchasing this book!

Do you like to read on the go but are unable to carry your print books everywhere?

Is your eBook purchase not compatible with the device of your choice?

Don't worry, now with every Packt book you get a DRM-free PDF version of that book at no cost.

Read anywhere, any place, on any device. Search, copy, and paste code from your favorite technical books directly into your application.

The perks don't stop there, you can get exclusive access to discounts, newsletters, and great free content in your inbox daily

Follow these simple steps to get the benefits:

1. Scan the QR code or visit the link below

https://packt.link/free-ebook/9781804618127

2. Submit your proof of purchase
3. That's it! We'll send your free PDF and other benefits to your email directly